Monographs of the

Rutgers Center of Alcohol Studies

No. 10

Monographs of the Rutgers Center of Alcohol Studies

Under the editorship of MARK KELLER

This monograph series was begun as Monographs of the Yale Center of Alcohol Studies and Numbers 1, 2 and 3 in the series were published at Yale. Beginning with Number 4 the series has been continued as Monographs of the Rutgers Center of Alcohol Studies. The change conforms with the transfer of the Center from Yale to Rutgers University. The works published in this series report the results of original research in any of the scientific disciplines, whether performed at Rutgers or elsewhere.

No. 1. Alcohol and the Jews; a Cultural Study of Drinking and Sobriety. By CHARLES R. SNYDER.

No. 2. Revolving Door; a Study of the Chronic Police Case Inebriate. By DAVID J. PITTMAN and C. WAYNE GORDON.

No. 3. Alcohol in Italian Culture; Food and Wine in Relation to Sobriety among Italians and Italian Americans. By GIORGIO LOLLI, EMIDIO SERIANNI, GRACE M. GOLDER and PIERPAOLO LUZZATTO-FEGIZ.

No. 4. Drinking among Teen-Agers; a Sociological Interpretation of Alcohol Use by High-School Students. By GEORGE L. MADDOX and BEVODE C. MCCALL.

No. 5. Drinking in French Culture. By ROLAND SADOUN, GIORGIO LOLLI and MILTON SILVERMAN.

No. 6. American Drinking Practices; a National Study of Drinking Behavior and Attitudes. By DON CAHALAN, IRA H. CISIN and HELEN M. CROSSLEY.

No. 7. Problem Drinking among American Men. By DON CAHALAN and ROBIN ROOM.

No. 8. Problem Drinkers Seeking Treatment. By EILEEN M. CORRIGAN.

No. 9. Drinking, Community and Civilization; the Account of a New Jersey Interview Study. By HAROLD FALLDING with the assistance of CAROL MILES.

No. 10. The African Beer Gardens of Bulawayo; Integrated Drinking in a Segregated Society. By HARRY F. WOLCOTT.

The African Beer Gardens of Bulawayo

Bonke abantwana abale minyaka enga pansi kwe 19 ubudala babo abala mvumo yokungena kulezi indawo. Umunti wonke ozakweqa lo umteto uzabotshwa.

All juveniles under the age of 19 years are prohibited from entering these premises. Any person who contravenes this order may be prosecuted.

The African Beer Gardens of Bulawayo

Integrated Drinking in a Segregated Society

BY

HARRY F. WOLCOTT

PUBLICATIONS DIVISION
RUTGERS CENTER OF ALCOHOL STUDIES
NEW BRUNSWICK NEW JERSEY

Library of Congress catalog card number: 73-620218
ISBN: 911290-42-7 ISSN: 0080-4983

MANUFACTURED IN THE UNITED STATES OF AMERICA

CONTENTS

Part III. Professional Involvement with Urban African Beer Drinking

Part IV. Individual Adaptation to Urban Drinking

List of Plates

Acknowledgments

Three institutions were instrumental in providing the resources for pursuing this study. The University of Oregon provided the opportunity of and support for a sabbatical study year in 1970–71, the year in which the fieldwork and the initial stages of writing were completed. Daystar Communications, with its African field office located in Bulawayo, provided a research stipend and other assistance during the year of fieldwork. Members of the Housing and Amenities Department (African Administration) of the City of Bulawayo offered not only the formal permission necessary to conduct the inquiry but extensive cooperation of their personnel, a generous sharing of facilities, and a sustaining interest in the completed project.

Personal acknowledgments are noted throughout the text. I have assumed that certain people who rendered opinions about facets of urban African drinking might prefer to remain anonymous, and if I have erred in failing to credit my sources it is because I opted for what seemed the lesser error between alternatives. In initiating the study, the cooperation of Eric Van der Meulen, the encouragement of Marshall Murphree and Donald K. Smith, the climate of openness to inquiry created by E. H. Ashton, and the rapidly expanding data base accumulating under the direction of Joan May and D. H. Reader were crucial. The continued and patient help of James Howard was essential and evident throughout the fieldwork. Ronnie Anderson, Roger Barber and Michael Ndubiwa were particularly attentive to helping me get my facts straight. For conducting the survey of beer-garden patrons I had technical assistance from Gordon Mullenix and extensive help with interviewing and basic tabulations from Robert H. Mpofu and Eliah P. Mupedziswa. Norman Delue, E. C. Greenfield, James Howard, Joan May and Eliah Mupedziswa read the first draft; Michael Ndubiwa read that and each subsequent one as well. Their suggestions led to substantial changes and refinements in the course of the fieldwork and in subsequent revisions of the written statement. A second draft was read and discussed by several of those mentioned above in a series of three staff seminars conducted at the office of the Housing and Amenities Department just prior to my return to the U.S.A. Professors Bryce Johnson, Clive Kileff, Marshall Murphree and Alfred G. Smith provided the invaluable service of collegial review.

H. F. W.

Eugene, Oregon
November 1973

Foreword

THE FIELD OF ALCOHOL STUDIES will benefit in many ways from this excellent study by Professor Wolcott. To understand the complex phenomena associated with the use of alcohol the field clearly lacks more cross-cultural data, particularly from Africa. We need to know more about the meaning of alcohol in various societies, the relationship of alcohol use to alcohol misuse, and the relationship of drinking to other activities. It is my opinion that Professor Wolcott adds much to what little is known in these areas. His work is in keeping with the high quality of previous monographs in this excellent series produced by the Rutgers Center of Alcohol Studies.

This monograph is scholarly, comprehensive and lucid in its depiction of the role of alcohol in African society. Further, by taking alcohol use as his departure point, Professor Wolcott is able to examine a surprisingly large number of other factors affecting life in Bulawayo, as perhaps only an anthropologist can do. Among the phenomena examined are Black–White relationships, alcohol and economic factors, alcohol in its relationship to culture change, and the focus of the public on beer garden problems as a surrogate for more serious social and political problems.

Moreover, Professor Wolcott produces a number of excellent insights which have relevance for the understanding of alcohol use elsewhere. At the risk of stealing his thunder, I would like to offer a few examples which I found particularly illuminating. For example in Chapter 4, Professor Wolcott discusses at length the socially integrative functions of beer drinking for urban Africans, as well as other positive functions. He writes (*p. 96*):

> "Perhaps the beer gardens provided the African in the urban setting a down payment on the promise of the city, a compensation for the growing sense of isolation associated with modern industrial societies and the accompanying anxieties and frustrations of putting one's fate in the hands of strangers. True, for too many the down payment of happy hours spent with colleagues in the beer gardens may have been one of the few rewards the city had to offer, but at least it provided a setting where one could revel in the freedom of urban autonomy and the euphoria of temporarily forgetting his woes."

He also makes the excellent point that while drinking may be integrative for the social group, at the same time it can be disintegrative for the individual (*p. 175*). This is the crux of the problem, not only for the African societies described by Professor Wolcott, but for many other societies as well. Thus, in the United States individual drunkenness and alcoholism are the high price paid for the socially positive functions of alcohol.

When viewing certain social groups—those at the bottom of the hierarchy—it is easy for other social groups—those at the top of the hierarchy—to forget the positive functions of alcohol and to focus on the damaging aspects. Professor Wolcott notes official White concern, as well as the concern of numerous Africans that the drinking of beer is damaging to the individuals in two primary ways: (*1*) it represents misspent money that could rationally be used to improve diet or living conditions; and (*2*) it sometimes leads to addictive drinking or anti-social behavior. The same observations have been made in the United States vis-à-vis Indians, Blacks, Chicanos, Eskimos and other oppressed minority groups. Why, indeed, do people seem to act against their own best interests? My own experiences in Africa 15 years ago raised similar questions.

During the late 1950s, I lived among the Zulu in South Africa. Half my time was spent in a rural cattle-keeping area in the foothills of the Drakensberg mountains, and the other half in the beautiful resort city of Durban on the Indian Ocean. In reading Professor Wolcott's fine book on the Ndebele and Shona of Rhodesia, I was constantly reminded of my own experiences and observations among the Zulu. This is not surprising since there are powerful and striking similarities between the Africans we each studied (all Bantu speaking, with similar cultures) and the Whites (colonial types, British and Africaaners) and particularly in the political situation which can only inadequately be described as "benevolent" oppression.

As I, too, alternated my time separately amongst the Africans and Europeans, I came to know much concerning what each thought about the other. Like Professor Wolcott, I noted a peculiar situation: the White political domination of the Blacks was virtually total, such that opportunities for social mobility were nonexistent. Living conditions, opportunities for work, pay scales and the like for Africans were determined by the Whites. Having thus set the limits of the conditions under which the Africans could live, many Whites then drew some perverse sense of satisfaction by pointing out that the way Africans lived was "primitive" or "childlike" or "immature." The Africans were not considered capable of "civilization," of governing themselves, of acting rationally.

Among the two most "damaging" observations of Blacks made by Whites was the fact that, although African incomes were extremely low, considerable sums were spent on commercially produced African beer (*tshwala*) and gambling on horse races. It must be pointed out that both these activities were not only officially sanctioned by the White government, but that opportunities for both were provided and encouraged by the government. Thus, the beer gardens of South Africa, sponsored and regulated by the White administration, bore considerable resemblance to those described by Professor Wolcott in Bulawayo—though the Bulawayan beer gardens appear considerably more attractive. Similarly, the race tracks in Durban and Johannesburg provided "betting rings" for Asians and Africans.

"Why do the Africans drink and gamble?" I was asked over and over. "They claim they are underpaid to the extent that they do not have enough to live on—and yet one sees them at the races and beer gardens." Professor Wolcott takes up the issue of why Africans drink and writes eloquently (*p. 75*):

> "The motivations of the urban drinker had also changed. If traditionally he drank in celebration, now he found less occasion to celebrate. His work was never 'done' in the shop or factory unless he had been dismissed or laid off, in which case he had little cause to celebrate as he joined the masses of the unemployed. The receipt of his weekly or monthly pay envelope marked only a momentary respite in his unending round of immediate expenses, and the amount of his pay, contrasted with the costs of rent, maintaining a family, paying off loans, meeting school fees, and so forth, served to emphasize the near futility of his endeavor. His industrial circumstances offered no possibility for the windfall of an unexpected 'good' year, the one sustaining hope in an agrarian lifestyle."

Thus, in Professor Wolcott's view, drinking now served as a release from overwhelming personal and social problems. Over 35 years ago, George Orwell, in his lesser known classic *The Road to Wigan Pier*,[1] also took up the issue of why the hopelessly poor of England used their inadequate income unwisely, sometimes to gamble on football pools, sometimes to spend on beer, other times to spend on

[1]ORWELL, G., The road to Wigan pier. (Berkley Medallion Edition.) New York; Berkley; 1967. [Orig. 1937.]

sugar for their tea, rather than on foods with reasonable nutritional value.

Orwell pointed out that in times of extreme poverty, such as during the depression he was writing about, people try to make things tolerable by lowering their living standards. He wrote (*p. 83*):

"But they don't necessarily lower their standards by cutting out luxuries and concentrating on necessities; more often it is the other way about—the more natural way, if you come to think of it. . . . and above all there is gambling, the cheapest of all luxuries. Even people on the verge of starvation can buy a few days hope ('something to live for,' as they call it) by having a penny on a sweepstake."

Later he wrote (*p. 88*):

"The basis of their diet, therefore, is white bread and margarine, [tinned] corned beef, sugared tea and potatoes—an appalling diet. Would it not be better if they spent more money on wholesome things like oranges and wholemeal bread or if they even saved on fuel and ate their carrots raw? Yes, it would, but the point is that no ordinary human being is ever going to do such a thing. The ordinary being would sooner starve than live on brown bread and raw carrots. And the peculiar evil is this, that the less money you have, the less inclined you feel to spend it on wholesome food. . . . When you are unemployed, which is to say, when you are underfed, harrassed, bored, and miserable, you don't want to eat dull wholesome food. You want something a little bit 'tasty.' There is always some cheaply pleasant thing to tempt you."

Obviously, there is more to self-destructive drinking and gambling than is satisfied by Orwell's explanation. But the problems with explanations offered by experts in the field of alcoholism are similar to those which Professor Wolcott observes regarding professional views in Bulawayo. On pages 169–170 he writes:

". . . . people's roles vis-à-vis both each other and urban drinking seen as an institution seemed to provide a reliable guide for predicting their behavior, their beliefs and the aspects of drinking that they would emphasize because of its relevance to their own role. The wary and concerned took their stand on drinking in terms of roles they had accepted in society; these roles enabled them to perceive problems selectively. There seemed to be no surprises. With rather good accuracy we might have predicted the aspects of urban drinking most likely to be singled out by missionaries, medical doctors, social workers or middle managers like the Area Controllers or bar managers.

"Collective comments and criticisms about urban drinking tended to be ritualistic in nature. Each individual's role prescribed his basic stance toward the problems, and his comments dealt with ritual observations and complaints in a manner more closely identified with his colleagues-in-role than in the complexity of the problems at hand. Thus each outsider addressed himself to that part of the problem, or that part of the African, that related to his own sphere of influence. Doctors worried about their patients' livers but had only fleeting information about their patients' lives: I recall one ward tour in which my medical host was honed in so thoroughly to a number of cirrhotic livers among recent hospital admissions that he probably never realized that he failed to address his patients as he guided my attention from one liver to the next. Missionaries expressed concern for the hereafter awaiting their potential converts but seemingly failed to appreciate that drinking beer might provide a welcome respite in the drinker's present circumstances. The police played off the realities of their work against visions of Utopia, presumably ignoring that job opportunities in their chosen line of work might be nonexistent were it to become a reality."

These views and those of professionals elsewhere tend to be atomistic. Among Western men of science, views of alcoholism are also dominated by professionalism and in ways even more atomistic than Professor Wolcott found in Bulawayo. Thus in the United States, even the concerns of medical specialists tend to reflect their expertise: liver, stomach, brain and psychiatric. And this is in addition to the legal, psychological, sociological, political, religious and other special-interest professions also involved in the problems associated with alcohol use.

A final point made by Professor Wolcott deserves some elaboration. He discusses how attention to superficial problems associated with beer production, quality control, alcohol content of the beer, proper distribution and the like tend to place more serious social and economic problems in the background. He writes (*p.* 229):

"It was certainly right to see attention given to such details as the nutritive value of the beer or the number of benches in the beer gardens, but it was unnerving to realize how seldom anyone's identification of critical social problems went beyond the mundane. It was not the selection of topics that seemed wrong, but the emphasis which various topics received. The emphasis on accountability and administrative detail tended to stabilize and sanctify the existing status quo. African discontent was funneled into acceptable channels—a man would tell you he would like to see more beer mugs or benches but dared not elaborate on the comparable shortage of wages, jobs, or political representation."

This important point is often true about many public issues in complex, multiracial societies, namely, that the politically dominant group will permit the "submissive" group to criticize the status quo, but only on banal matters of form and rarely on critical, basic, structural issues. Given the climate of opinion is which he has worked, Dr. Ashton's straightforward Introduction to the study, written from the perspective of the on-the-scene administrator, reveals something of the ambivalence and tension felt by at least a few members of the dominant group, and the accomplishments of his department may reflect his own clear sense of purpose regarding a course of action.

And extending Professor Wolcott's point regarding surrogate issues in the area of professional concern surrounding alcohol use, in the United States one can see how issues are shifted concordant with the views of members of the alcohol studies establishment. Here I am referring to what often appears as the consistent attempt to place the "blame" for the social problem of alcohol on drinkers and alcoholics, rather than on the social and cultural forces that lead, indeed almost coerce, people to drink. Similarly, if alcoholics fail to recover, the blame is theirs for "improper socialization" or "insufficient motivation," rather than on the therapist for insufficient skill, or the health care system for its failure to provide realistic and workable treatment programs, especially for the more difficult and harder to reach problem drinkers. Thus, in the United States at least, the person who suffers most from alcohol is the person with the fewest resources for change and is the one charged most with the responsibility for changing behavior.

Here we may paraphrase MacAndrew and Edgerton[2] who wrote *(p. 173)*, "Since societies, like individuals, get the sorts of drunken comportment they allow, they deserve what they get." Those societies which encourage unlimited, unrestricted drinking, and which fail to provide alternatives for the functions fulfilled by alcohol, and which further fail to develop adequate means to deal humanely with individual problem drinkers, and, finally, which fail to develop programs which would minimize the socially disruptive aspects of alcohol use, deserve the alcohol problems they get.

Norman A. Scotch

Department of Socio-Medical Sciences
Boston University School of Medicine

8 February 1974

[2]MacAndrew, C. and Edgerton, R. B. Drunken comportment. Chicago; Aldine; 1969.

Introduction

I am proud and delighted that Professor Wolcott should have asked me to write this introduction to his monograph. For he is an outstanding American anthropologist and one who has dared to work in a difficult, controversial field. As a scientist, he has followed the truth wherever it led, objectively and courageously, and as a humanist he has shown great understanding and sensitivity in presenting his findings delicately but none the less forthrightly, even in areas where they could hurt. In doing so, he has made a fine contribution to the study of urbanization in developing countries, and a fascinating analysis of an important and involved urban institution.

An invitation such as this is rare if not unique. It is as if a microbe hunter asked his quarry to chase him in turn. Very few subjects of anthropological inquiry are sophisticated enough to cross-examine the inquirer, and even fewer inquirers would welcome being subjected to such an examination. Professor Wolcott is the exception and I welcome his initiative even though I will do no more than set the scene of his research. I also appreciate his giving me another rare experience — the experience of being a researcher at the receiving end of a research project; I have been an anthropologist myself and have turned the searchlight on others. Now the searchlight was turned on me. It was an embarassing and salutary experience. Let me briefly explain how all this happened.

I am a Southern African through and through. I was born in Lesotho and at one stage spoke better Sotho than English. As a postgraduate I did field research among the Basotho, and for many years kept in touch with those turbulent, delightful people. I was also for a short time associated with the British Government's official inquiry into ritual murder, an endemic problem that in the 1940s assumed epidemic proportions. I became a traditional colonial administrator and served in Botswana where I did some further research both on my own and together with that eminent Africanist Professor I. Schapera. Then, like so many of my compatriots, I drifted to the towns. After 3 years in Johannesburg I came to Bulawayo in 1949 to establish a new department of urban African administration.

The basic aim of the department was, and still is, to eliminate and prevent slums, those social horrors that are the usual concomitant of urban growth. This is an unending, unrelenting assignment, and shares the plight of the Red Queen in Alice's Wonderland that "It takes all the running you can do just to stay in the same place. If you want to get somewhere else you must run at least twice as fast as that." To let up for a moment is to court disaster. The African population has trebled in 2 decades to over ¼ million, and is expected to double again in the next 20 years. This rapid growth is typical of the world-wide phenomenon of urban development and the cityward migration of workers followed by their women and children. Fortunately we have managed to contain this migration within manageable proportions, and though we have not entirely eliminated its usual concomitants of shantytowns and slums, we have kept them to a minimum.

The human problems arising from this movement are well known and are basically much the same here as elsewhere. The majority of those who come to town are peasants, used to a rural atmosphere and way of life. Faced with the competitiveness and increased tempo of town, with its different standards and different demands, they have to adjust or return to the country, and those that survive find themselves becoming urbanized and detribalized. In addition, an increasing proportion of townsfolk either came as children or were born here, and so are even more urbanized and less tribalized than their parents, thus adding another dimension to the urban problem. There is yet another factor, the racial one, and this is a painful one we share with other multiracial areas such as the United States of America. The towns are "White" whereas the immigrants are "Black." The towns are White in that the enterprise that stimulates their growth and attracts the immigrants is White, the wealth is White, the control and structure of the city is White. But in North America, "White" relates mainly to race or color, since the Blacks speak the same language as the Whites, have the same values, ambitions, customs and mores; whereas in Southern Africa the position is more complex. Here, White refers to cultural difference as well, for White also means western European; the language is English, modes of thought, customs, behavior, religion and economics are Western, Christian, highly individualistic and competitive. So our immigrants are not only being urbanized and detribalized, they are also becoming Westernized.

The objective of my department is to help these sundry processes of change along as smoothly and painlessly as possible.

The first task is obviously housing. Thatched mud huts are all very well in scattered rural villages, but they clearly won't do in the closely packed conditions of a city. As the peasant immigrant has neither the resources nor the know-how to build for himself in town, he is dependent on private enterprise or public authority to do it for him. As the former tends to be exploitative, Rhodesia has gone for the public housing authority. This is where we come in. We lay out housing estates, provide the basic services of roads, water, light and sewerage, and build the houses for letting or purchase. Latterly some of the immigrants have become urbanized and affluent enough to start building their own houses, using the normal contract sources for finance and construction.

The second task is to help turn houses into homes. This is being tackled in a host of different ways, at different levels of sophistication. They include family casework, the encouragement of women's, girls', and boys' teen-age clubs, the formation of cultural guilds, preschool centers and the provision of swimming baths, stadia, sports fields, parks and playgrounds, libraries, art centers, from the purely recreational to the instructional, such as cooking, dressmaking and budgeting. Many are designed to help people from the simple subsistence economy of the country where many things — water, sanitation, fuel — are free or are provided by the family from its agricultural or primitive food gathering, to the cash economy of town, where practically everything has to be paid for. Unlimited leisure time is no longer the satisfying thing it used to be; regular employment becomes a necessity and unemployment a disaster. We are therefore having to place increasing emphasis on home industries and income-earning craft projects for school-leavers who cannot find places in the formal economy. Formal schooling and training is provided by Government and a few religious organizations, but there are gaps which we help to fill through adult literacy classes and the libraries. Above all, new social ties must be formed to strengthen or replace the traditional kinship bonds that have been weakened or broken by moving from village to town. Many of these new ties spring up spontaneously in answer to new needs, such as burial or mutual aid societies formed by kinsmen or fellow tribesmen from local areas in the homelands. Others have to be nurtured as they grapple with new concepts and problems — cultural societies, club organizations, preschool associations and civic boards. These groups are valuable in themselves, but all of them are important in discovering and nurturing new forms of leadership and experience appropriate to modern urban demands.

All this needs money; money for housing, money for buildings and equipment, money for staff. The third task of my department is to make this money, and this is done through a brewery of some magnitude, and a large liquor undertaking.

Nearly a century ago, by international agreement, the great powers, mindful of the devastation that had been wrought elsewhere by gin and brandy, undertook to prohibit the sale of western alcoholic beverages to the indigenous peoples of Southern Africa. Fortunately the latter already had satisfactory drinks of their own which could be easily brewed at home. In the countryside there was therefore no problem, but in the towns this was obviously not the case, and various forms of illicit liquor production and sale developed. So a system of municipal monopoly developed, whereby the municipalities could, if they wished, brew and sell the traditional African beer. Over 60 years ago, Bulawayo opted in favor of this and showed it could be made to pay dividends, both socially and financially. From this developed the concept that beer profit should be used exclusively for the benefit of the African section of the population, and eventually the corollary that activities for the benefit of the African population should be funded only from this source.

Although one can argue the moralities of this system, there is no doubt about its practical advantages. Granted that people will drink whether one likes it or not, it seems sensible to channel the profit into the public purse rather than a private pocket, and at the same time keep liquor consumption within bounds and avoid the excesses that come from private exploitation. We have tried to do this intelligently and to kill two birds with one stone, to provide a socially valuable form of recreation and to nurture a valuable source of income. So we have steadily expanded the municipal brewery, until its output is now one of the largest in Southern Africa, and is significant by ordinary commercial standards. This has gone hand in hand with the steady expansion of beer gardens, bars and bottle stores, and the development of a multimillion-dollar business with an enviably profitable return, every cent of which is ploughed back into housing, health and social provisions.

The ambivalence of my position is obvious. How can one maintain a healthy community and a healthy profit at one and the same time? I can almost hear a critical reader questioning my morality and even my sanity. And why not? I have often done so myself. My answer to my self-questioning has been favorable! What is the reader's? I accept it is unhealthy that the provision of social services should depend upon the size of the profit, but I feel we should be practical about it and recognize that much practical good can come out of a hypothetical evil. As an optimistic voyager in these dangerous seas, I have tried to steer between the Scylla of driving people to drink by aggressive salesmanship and the Charybdis of starving the social services through hypersensitivity to the questionable source of their finances. Thanks to ever readiness of critics of all persuasions and to the constructive comment of African and European civic leaders, I hope and believe I have been making a safe passage. But it is easy enough to delude oneself, and so when Harry Wolcott hove on my scene as a professor in search of a project, I jumped at this opportunity to have it critically and impartially surveyed.

Robert Burns, contemplating a posturing damsel unconscious of the louse on her hat, penned his immortal warning against complacency:

"O wad some Pow'r the giftie gie us
To see oursels as others see us.
It wad frae monie a blunder free us
An' foolish notion."

We heeded this warning and have greatly benefited from this unique opportunity and I hope will continue to do so. For Professor Wolcott has examined the beer garden as an institution and as a symbol of our dilemma from every angle and in great depth. He took considerable pains to check his facts and validate his conclusions, and in this showed a conscientiousness and a concern for objective accuracy that was an outstanding credit to his profession. In the course of this, he discussed his drafts with various people who could critically appraise his findings, including my senior staff and myself. I hope our discussions were helpful to him – the feedback was certainly invaluable to us. In some of our sessions we may unconsciously have tried to amend the last two lines of Burns' verse to

How much better if by some spells
Others could see us as we see oursels,

but he would have none of that. He recognized our problems but he stuck to the truth as he saw it, painful though it might be.

What of the future? I am not seer enough to look very far ahead. But one thing is certain, the African urban community is becoming more and more urbanized, more sophisticated, more articulate. Dilemmas such as I have described in the Introduction, and controversial institutions such as Professor Wolcott has analyzed in his monograph, are no longer the sole concern of administrators, but are becoming the concern and the responsibility of new leaders. Their initial simplistic reaction, "don't-build-any-more-beer-gardens," has changed to "don't-build-any-beer-gardens-where-anyone-could-see-them," and is now changing again to more realistic appreciation of the complexities of the problems involved—"let's-build-these-beer-gardens-with-these-and-these-safeguards." As Africans participate further in administering their affairs, still greater changes will take place. Whether they will be toward more widespread public ownership and control of the liquor business, or whether they will veer more to private enterprise as is so common in western practice, is anyone's guess—there are powerful conflicting forces favoring each alternative. There will undoubtedly be changes in drinking patterns and habits, and probably a greater proportion of bottled beer and spirits will be consumed at the expense of the traditional brew drunk in plastic mugs. But the beer garden will still be there, as gay, as lively and vital as ever, or town life will be the duller for its passing. What a fascinating field this will be for a latterday Wolcott, and what a wonderful starting point this monograph will give him for his comparative research!

E. H. Ashton

Housing and Amenities Department
City of Bulawayo, Rhodesia

17 December 1973

Part I

The Scene and the Study

1

Overview: Integrated Drinking in a Segregated Society

IN LETTERS from friends and colleagues during a year spent in Rhodesia I received a good deal of ribbing about making a study of urban African drinking. "What an interesting solution you found to your research problem: drinking!" quipped one associate. Another observed, "Wait until the legislators find out that they let one of their professors go off for a year of drinking with the natives." Actually, I did not go to Rhodesia for the express purpose of studying African beer drinking.

What I did set out to do was to learn something about contemporary African life by engaging in cross-cultural study in an urban setting. It was my intention to pursue the fieldwork approach common to the cultural anthropologist, to attempt to "live my way" into African social life. I anticipated little difficulty in identifying one or more problems for study relevant to my professional interests in education and anthropology, particularly along the lines of cultural transmission and culture change. During the course of early discussions with persons conversant with the Rhodesian scene, while still gathering some first impressions of facets of Rhodesian African life that might be personally and academically interesting, I identified 24 researchable topics. Urban beer drinking was the 24th. Had one of my earliest inquiries to an official in the Ministry of African Education been met with a favorable reply, I might have embarked along a quite different line of inquiry, probably delving into formal and informal aspects of enculturation among urban African youth.

What I had not fully comprehended before arriving in Rhodesia was that it would be virtually impossible for me to conduct research into African life in the casual, comfortable and familiar role of participant–observer, at least in the urban setting. In urban Rhodesia, I quickly learned, Whites lived in "European" areas and Blacks lived in "African" areas referred to as "townships" or collectively as "The Location." That ethnolocal residence pattern seemed "natural" to some and "inevitable" to most. I was free to come and go pretty much as I wished, but it was assumed that if one went into the African townships adjacent to any White urban area he conducted his business quickly and then departed. Few Whites ventured into the townships after dark; those who even discussed the possibility of doing so were of the general impression that at least technically they were required to register with "somebody," although there was some doubt about just which official it would be, and the question was essentially hypothetical, since the kind of people who might seek permission were not usually those who made the trip. It was possible for me as a social scientist to be both a participant and an observer in Rhodesia, but there was little ambiguity about which of those roles I was enacting at any given moment.

Only in the White community was I a participant. I slept, ate, banked, attended the theater, entertained and was entertained there. Concomitantly, in my excursions into the African community I was always and only an observer. That was the local custom, a vestige of a colonial past whose influence lingered on in the absence of any strong pressure to replace it. The best alternative for me was to bring my usually open-ended approach for inquiring into human social behavior into sharp focus, to look carefully at quite specific behavioral dimensions rather than to enjoy the luxury of the old-style fieldworker who attempts to take a long slow look at anything and everything of possible consequence in the unfamiliarity of a new cross-cultural setting. The urban African beer gardens, an institution provided for the indigenous population and restricted exclusively to their use, seemed a likely and interesting setting on which to focus an inquiry.

In a country externally harassed by political ostracism, unsympathetic review, and international economic sanctions against what was often referred to as "Prime Minister Smith's illegal regime," and internally obsessed with the eventuality of a political minority of 5,000,000 Africans (1970 population estimate) forcibly altering their position vis-à-vis a political majority of only 240,000 "Europeans," virtually every social problem was a sensitive one. Drinking, at least in terms of the disruptive social consequences which are customarily attributed to it—sexual promiscuity, aggravated assault, accident, unstable home life, juvenile drinking, alcoholism—is always a problem of community concern. In the Rhodesian setting the municipal governments of the larger communities not only accepted responsibility for confronting these problems among their rapidly expanding numbers of African urbanites but, as the brewers and retailers of beer produced especially for that population, appeared liable to the charge of causing the problems, or at least contributing to them, in the first place. "The officials have very tender consciences on this issue," I was advised.

Nevertheless, my probe into the possibility of learning about the beer gardens received the approval (and presumably even the blessing) of some of those very officials. Perhaps they recognized certain advantages in encouraging my inquiry. Three advantages spring easily to mind.

First, there was no question that beer drinking, and the municipality's role in it, raised a series of sensitive social issues. Critics of the beer gardens included not only their ritual adversaries (one missionary view was that the problem of African drinking was the preeminent problem of the times) and recurring outbursts from some sectors of the White community ("Are we encouraging a nation of alcoholics?") but also a growing concern among urban Africans themselves about the wisdom of making beer so readily accessible to their sons, brothers, husbands, and even their women.

Second, a study considerably more sociologically oriented than what I had proposed was already under way into urban African drinking patterns. The major locus of that study was in Salisbury, the capital city and home of the University of Rhodesia, under whose auspices the research was being conducted. Survey data paralleling that of the Salisbury study had also been collected from a sample of households in several of the Bulawayo townships, and plans were in the offing for exploring other avenues of related research. The initial reaction was that my interest might complement ongoing research and that perhaps I could participate as a team member in that work. This proposal aroused a suspicion on my part that by being assigned to a locally instigated research enterprise I was being relegated to a long and carefully guided tour so that I would eventually see the rationale for municipal policies. My suspicion was ill-founded.[1] Although direct costs of conducting the University's study were met by the municipalities, the research program operated rather inde-

[1]The very fact of being accepted as a researcher and of feeling that I was not being given a guided tour also influenced my perspective and the subsequent course of my fieldwork. I have discussed these "feedback influences" in a separate article which deals more fully with the methodology of this study (142).

pendently of them. The survey of African drinking patterns was housed in the Institute for Social Research of the Department of Sociology and was known as the Alcoholism Research Unit. At our first meeting, Mrs. Joan May, the Research Fellow supervising the project, laid out everything that had been done, provided a summary of the data from the Bulawayo survey, and loaned me a set of field notes collected during 51 open-ended interviews made in or about the beer gardens and liquor stores in Salisbury's townships. The work that had been done provided not only a solid research base on which to begin my inquiry but made it possible to draw on comparative data collected by other researchers (including predominantly African fieldworkers) in a different urban setting. I have made extensive use of data collected by the Alcoholism Research Unit and of insights suggested by persons associated with it.

Third, in light of the relevance of the topic as a sensitive social issue and the fact that further inquiry was not only solicited but had already been initiated, there was the potential advantage of having an outsider, free of commitments to local or national ties, lending some new perspective to both scope and method. There are obvious risks in this approach as well, depending both on the degree of objectivity attained by the researcher and the tolerance for objectivity on the part of others toward the issue at hand. I was told that I was indeed fortunate to have come to Bulawayo, for the personnel charged with administering the African townships there were recognized widely and among diverse groups for being exceptionally capable and enlightened, and Bulawayo's townships were, with allowance for the excesses of local pride, generally considered to be "as good as or better than" those of any other local government in southern Africa.

From a personal point of view, there were still further advantages in directing my attention to urban drinking. First, the topic was focused and manageable within limitations of time, resources and concurrent commitments. Second, a study of drinking behavior and attitudes about drinking provided a good vehicle for studying and comparing cultures. Throughout most times and places human beings have brewed and poured mildly intoxicating libations and have designated the occasions, the statuses, and the amounts appropriate for consuming them. The custom of preparing and drinking beer is so widespread that some anthropologists have suggested that it provides a more universal aspect of culture than does the preparation of bread, and they have pondered the somewhat remote possibility that "thirst rather than hunger may have been the stimulus behind the origin of small grain agriculture" (9, *p. 516*). From growing the ingredients necessary for brewing to distributing and consuming the finished product, beer tugs variously at the web of culture regarding the division of labor, the reckoning of time and season, sharing and hoarding behavior, the validation of events worthy of collective celebration or grief, and the mechanisms of social control. For the behavior of the drinker himself, the effects of beer in stimulating expression and lowering inhibition reveal the depth of the imprinting of culture, for the constraints upon the behavior of a man who is "so drunk he does not know what he is doing" provide ample testimony to the fact that drinking behavior is both manifested and viewed within the context of cultural norms (81). Attitudes related to drinking behavior also provided me with a way of looking at some of the pervasive problems confronting urban Africans—particularly those related to opportunity for employment, formal education and other routes of access to the life-style of the dominant culture—from the human perspective of particular individuals rather than from the organizational perspective of the formal institutions themselves.

The present study is intended to be descriptive, to provide evidence of the complexity and context of urban African drinking rather than to offer answers to questions that have not been formulated. It suffers from a limitation that bothers an anthropologically oriented researcher more than it might bother social scientists from other disciplines. I had to rely to a great extent on what people told me, rather than on what I could observe firsthand or fully substantiate with subsequent observations. In addition to my opportunities for observing and interviewing during my 12-month stay, I have relied on interview data collected by others under circumstances not totally apparent in the protocols themselves. I have relied on what officials have told me, never quite knowing what information they might be intentionally avoiding or unintentionally omitting. And I had to rely on whatever comments African respondents and informants chose to volunteer to a White stranger in their midst, a situation not only rife with political overtones and the almost inevitable cross-cultural suspicion and wariness but also a highly unusual event for many of the younger Africans who reported having virtually no occasion or opportunity to meet with a White to converse about anything!

At the outset, a White Rhodesian asked me how I intended to dress for my forays into the beer gardens so as not to look too conspicuous. The question highlights the dilemma for the researcher who is accustomed to thinking of himself as a subtle research instrument. In a community with an African population of about two hundred thousand, it did not seem to matter a great deal whether a White observer wore a swimsuit or white tie and tails while he was carrying out his fieldwork. This is not to imply that I did not attend to what I wore. If I was accompanying European staff in their official rounds of the beer gardens, I donned my safari suit similar to the suits they wore so regularly that they had become an unofficial field uniform. When I was going to have the opportunity to talk at length with Africans, I wore shirt and tie in the custom of many urbanized Africans, especially those desirous of showing by their dress that they do not have to work with their hands. It was usually to my advantage to wear the bright shirts and broad ties not yet locally in vogue which, along with my American accent, I counted on to help relay the message I wanted to convey in the course of my fieldwork: I was an outsider, not part of the immediate problem of Rhodesia's Africans and, except for whatever contribution this study might make toward local and international understanding, not part of the solution.

Integrated Drinking: A Functional Perspective

Anthropologists interested in "alcohol and culture" have ordinarily contributed to one of two complementary aspects of the subject in their studies. First is the ethnographic reporting of drinking behavior in a particular society, such as the account provided here. The second is the comparison of aspects of drinking behavior across many societies in the attempt to identify transcultural variables or to test concepts and theories suggested by single cases.

A number of references to observations by anthropologists elsewhere, comparing or contrasting with drinking practices observed in Rhodesia, appear throughout the text. Growing out of such studies[2] and comparisons made among drinking practices in many societies are a number of hypotheses about the relationship between drinking and society. The following list presents briefly, and in the chronological order in which they were originally proposed, six

[2] See Popham and Yawney (98) for a bibliography.

hypotheses frequently cited in anthropological discussions about drinking:

1. Horton's hypothesis that the primary function of alcoholic beverages in all societies is the reduction of anxiety (56).

2. Lemert's hypothesis that drinking serves as a vehicle of social integration and a symbol of social solidarity (72).

3. Lemert's alternative hypothesis that drinking is an institutionalized behavior pattern operating relatively autonomously and is only tenuously related to other aspects of culture (73). That is, drinking and drunkenness may be expressions not only of culture patterns but of collective behavior, symbolic protest, individual demoralization, situational controls, and social interaction (76).

4. Field's hypothesis that excessive drinking is determined by the absence of corporate kin groups characterized by stability, permanence, formal structure, and well-defined functions which provide controls over heavy drinking that are not available to societies with informal or less well defined kinship systems (31).

5. Child, Bacon and Barry's hypothesis that drinking or drunkenness or both are frequent in preliterate societies which produce a high degree of conflict about dependence and independence (25).

6. Graves' hypothesis, dealing specifically with North American Indian migrants to the city, that migrants who have a personal goal structure least in line with urban opportunities will experience the greatest psychological conflict about remaining in the city and will therefore drink the heaviest (40, 41).

Interest in cross-cultural studies of drinking has paralleled the development of such studies of many other aspects of social behavior. The cross-cultural method serves a dual function: it not only provides a basis for testing hypotheses growing out of case studies but also a check against unconscious bias and a test of the generality of assumptions and findings. One of the special problems for the Western observer studying drinking is that within his own culture drinking tends to be an emotionally loaded issue. As Margaret Bacon has observed in discussing the contribution of the cross-cultural method in drinking studies (2), the tendency to assume that a high rate of alcohol consumption is essentially and conspicuously disruptive of social life, or to assume that drunkenness is inevitably a problem, appears to reflect beliefs generally accepted in our society that are not borne out in cross-cultural study. Indeed, at least one anthropologist suggests that American attitudes about drinking are particularly fraught with "conflict and ambivalence" even by comparison with standards prevailing in other Western countries. John Honigmann made the following comments comparing American drinking norms with those of an Austrian village (53, *p. 158*):

> "To a visitor from the United States, where alcohol is treated with nearly the same caution as guns, drugs, and germs, Western Europeans are amazingly casual about drinking. The American realizes by comparison how his land is unable to accept drinking easily, but instead surrounds it with restrictions, curtails it by difficulties, and conceals it from public view. Europeans' far more accepting attitudes are indicated by sidewalk cafés, by beer hawked through trains, by the beer and other alcoholic beverages sold from stands at railroad stations, and by the ease with which strong drinks can be obtained even in places that readily admit children. These traits do not suggest much conflict or ambivalence over the use of alcohol."

In moving from single cases to identification of transcultural variables, it is necessary to identify variables in such a way that their presence or absence in a variety of cultural settings can be reliably determined from a careful reading of ethnographic data. To enable one to make comparisons that are meaningful and manageable, it is also essential to reduce the potential variation among single dimensions by identifying clusters of variables that share some unity. These clusters provide themes for comparing related sets of behavior in different societies. Since Horton's study (56) in 1943, the last three decades have seen several major cross-cultural inquiries and publications providing a comparative basis for alcohol studies, such as Washburne (134), Field (31), Child et al. (25), Barry (4), and Bacon (2).

Particularly noteworthy is the comprehensive cross-cultural study by Child, Bacon and Barry, which appeared in 1965 as a Supplement of the QUARTERLY JOURNAL OF STUDIES ON ALCOHOL (25). The authors of that study used a sample of 121 preliterate cultures, in which drinking had been reported, to investigate a host of measures relating to the consumption of alcohol. Through factor analysis, interrelated variables were summarized into a number of independent factors. Four such factors were ultimately identified: *Integrated Drinking; Quantity; Inebriety;* and *Hostility.* It is the first factor, Integrated Drinking, which is of particular interest here. I have taken this factor suggested by cross-cultural analyses and used it as a point of departure for analyzing drinking behavior in a single case. It will be important to keep in mind, however, that drawing upon the concept of Integrated Drinking as a functional perspective for examining drinking behavior is not the same as judging drinking to *be* integrated in a given society or assessing the extent to which drinking in that society is more or less integrated in comparison with other societies.[3]

As used here, the concept of Integrated Drinking has been extended beyond the specific set of variables of which it consists as a statistical measure in drinking studies. "Integration" is viewed in a broader anthropological sense as the process of adjustments between the elements which comprise a culture (39, *p. 278*; 78, *p. 343*). Thus *drinking will be considered to be integrated in any way that it functions to enhance or maximize the survival chances of the culture and of the society that practices it.* Nevertheless, the specific behavioral variables clustered together in the Integrated Drinking factor described by Child, Bacon and Barry serve as a useful starting point for examining the concept. They identified seven such variables (25, *pp. 11-16*):

1. Frequency of drinking as a *religious* ritual: the act of drinking must have a supernatural purpose or meaning, be performed in the presence of ("real" or "imaginary") others who concur as to its meaning, be performed in a manner specified by social tradition, and be expressive and communicative rather than procurative.

2. Frequency of *ceremonial* drinking: alcohol is consumed in situations which are not strictly religious ritual but are nonetheless located in the context of, or ancillary to, culturally standardized recurrent rites, ceremonies, celebrations, or assemblies of general social significance and of a public or quasi-public nature (e.g., fiestas, holidays, victory celebrations, rites of passage), or family ceremonials of clear public import such as name giving, weddings, funerals, work parties, or prestige feasts given by an individual.

3. Extent of *ritualization*: drinking is surrounded by ritual restrictions as to time, place, material apparatus, manner of drinking.

4. *Quantity* of ceremonial drinking: the scale runs from "very small" (0.6 oz of alcohol, the equivalent of a 1½-oz "shot" of distilled spirits, a

[3]One dimension which can be used to sort the many cross-cultural studies on drinking is to contrast those which emphasize the socially integrative functions of alcohol and those which emphasize the socially divisive outcomes of drinking. Heath (49) provides a number of references illustrating each type of study. In the former category he includes Bruun (11), Field (31), Honigmann (53), Mangin (84), Netting (91), Rubington (109), Simmons (119) and Snyder (123). In the latter category he cites Bunzel (21), Hawthorn, Belshaw and Jamieson (48), Horton (56), Hurt and Brown (60) and Sariola (113).

bottle or can of "European" beer, or a small glass of wine) to "extremely large," 10 times that amount.

5. Quantity of drinking as religious ritual: the amount consumed on a *single occasion* during a religious ritual.

6. *Approval* of drinking: the degree to which majority opinion in the society approves of the sort of drinking typically found in the society.

7. *Extent* of drinking: an "across-the-board" estimate of the proportion of adults in the population who drink, regardless of quantity, frequency, or other qualifications.

Child, Bacon and Barry state that they chose the label "Integrated Drinking" because the pattern suggested by the first five of their seven factors shows alcohol use integrated with ceremonial and other occasions. What the variables show in common, they note, is "the use of alcoholic beverages in a context of positive social meaning" (*p. 13*). To provide an illustration of a society high on the Integrated Drinking factor, they give a brief account of drinking among the Ainu of northern Japan (*pp. 15-16*):

"Accounts of the use of beverages among the Ainu are fairly full, and . . . the alliance between drinking and social ceremonies is the most conspicuous theme in this material. The first tattooing of a young girl, weddings, housewarmings, funerals, anniversaries of funerals—all such ceremonies include the drinking of alcoholic beverages. The quantity typically consumed by each man participating is clearly very considerable. Some ethnographers speak of the ceremonies as little more than a pretext for drinking. It seems clear from their accounts, however, that the drinking serves as a powerful incentive for attending these ceremonies and thus indirectly serves an important integrative function. Among the ceremonial uses of alcoholic beverages are specifically religious ones. Libations are offered to spirits, often at the sacred spot within each family's hut. Drinking also plays an important part in ancestor worship. . . .

"There is considerable evidence of ritualization, going far beyond the fact that alcoholic beverages are used on ritual occasions. . . . It is clear that drunkenness is warmly anticipated and welcomed; and in all the material describing the reactions of Ainu to alcoholic beverages, reserve is indicated only in girls and young women, who are said to try to avoid drinking for fear of making themselves ridiculous and jeopardizing their good name."

In the pages that follow, I make full use of Integrated Drinking as a perspective for looking at the drinking behavior of Rhodesian Africans. The terms employed to describe variables in the cross-cultural study of drinking are all relevant in the African setting, and Integrated Drinking among the Ainu of northern Japan bears some remarkable similarities to African drinking behavior in Rhodesia.[4] In the latter setting, beer has always provided a material link between present generations and past ones, between the earthly and the ethereal. The ceremonial brewing and drinking of beer is an integral part of African life for occasions ranging from marking an important rite of passage to repairing the fences around one's kraal. Ritualization in the process occurs in proper ceremonial form from the very moment at which brewing begins, as the brewer addresses the ancestors and shows them the grain she intends to use to brew beer in their honor. Beer brewed for ceremonial purposes is ritually offered, ritually spilled, and ritually drunk. But even beer drunk at the urban beer gardens is drunk ritually as the large mugs are passed from one drinker to the next or momentarily set down while conversations or greetings dominate "drinking" behavior.

Large quantities of beer may be consumed—as a matter of fact, the amount of beer available was the critical element in determining how long the ceremony continued. Beer drinking was "approved" in the society as a normal part of human behavior; indeed, African beer has traditionally been considered more a source of nourishment than an intoxicant. The proportion ("extent") of drinkers among adults was undoubtedly high—perhaps it included all adults, men and women alike, although I found some variation in the opinions offered by informants. It is important to realize the significance of defining "adult" by the customs of the society under consideration, as Child, Bacon and Barry did in their rating on this variable (25, *p. 14*). Even when they reach the age of young manhood, African males traditionally have not been recognized as adults. My informants suggested that "in the old days" a man might not start drinking until late in his 20s or well into the 30s, an obvious point of contention in the urban setting where drinking is regulated by laws designed to prohibit alcohol use by children and juveniles rather than to recognize social maturity.

Two additional characteristics of societies rated high on the Integrated Drinking factor in cross-cultural study are of special interest here. The first is that such societies are generally able to tolerate drunkenness (i.e., they are rated high on the variable Approval

[4]More recent accounts suggest that drinking among the Ainu is no longer as well integrated as reported earlier. See, for example, Geiser (36).

of Drunkenness), a finding which runs counter to the expectation that the only way to accommodate the potential for social disruption is to disapprove of drinking in excess. This does not deny that relative sobriety may indeed characterize a society that shows Integrated Drinking—as illustrated, for example, in Snyder's study of Jewish drinking (123)—but Integrated Drinking is not predicated upon low consumption or disapproval of drunkenness. Although a wide range of attitudes about drunkenness expressed by urban Africans will be reported here, it is probably fair to say that as the drinkers themselves regard it, "being drunk" is neither reprehensible nor undesirable.

A second characteristic of societies rated high on Integrated Drinking is that they tend to be groups accustomed to drinking throughout the course of their own remembered history rather than those whose drinking began "after contact." This sense of continuity with the past is an important aspect of contemporary African drinking. This continuity contrasts markedly with the postcontact drinking of the Kwakiutl Indians of British Columbia, among whom I have previously had the opportunity of doing fieldwork, and with drinking among North American Indians in general. As a matter of fact, in spite of socially integrating functions that alcohol may perform in facilitating interaction among the Kwakiutl, I am inclined to think that the drinking behavior I observed in that setting is essentially disruptive in comparison with African drinking. It is interesting to note the absence of many of the factors formally identified with Integrated Drinking in the cross-cultural study in Kwakiutl drinking: it is not part of religious ritual, it is not part of ceremonial behavior, and it is not ritually used. Instead, the specter of participants or spectators drinking too heavily and "spoiling everything" looms at every Kwakiutl gathering, whether it is a school Christmas concert or the contemporary form of a potlatch ceremony. Elders beg and cajole their juniors to forego their informal drinking parties in the interest of maintaining collective solidarity at large social gatherings. Shortly after arriving in Rhodesia, I had occasion to write my old Kwakiutl friend Henry Bell, a former village chief who had long expressed his concern about the disruptive influence drinking was having on members of his family, his village and his people generally. I told him of my intention to study drinking behavior in an urban African setting. He wrote in reply: "Everything here about the same as ever. Lots of drinking going on. You have to tell me about the

Black people, if they doing any drinking as good as we are here."

Growing out of and extending beyond Integrated Drinking as a cluster of variables identified for a study in the cross-cultural method, I shall employ the idea of alcohol use in the context of *positive social meaning* as a perspective for discussing and analyzing the social matrix in which the African beer gardens of Bulawayo are embedded. This is not to judge drinking in the Bulawayo setting as "integrated," but to approach the ethnographic data with certain recurring questions relating to cultural integration: In what ways is any one aspect of urban drinking integrated with other aspects of African life? In what ways does the provision of beer by the European population for the African population act as a link between cultures, serving to integrate their commonality of purpose even as it maintains the dramatic distinctions which separate them? Or, in short, how do the urban beer gardens function?

I am not so committed to a functional perspective that I intend to ignore that a complex institution such as a municipal brewing industry may be expected to show some dysfunctional qualities as well. I shall identify these dysfunctional qualities as observed and will point out certain aspects of urban drinking and of a municipally sponsored liquor industry which appear to contribute to individual stress and collective disruption.

Perhaps individual biases need to be explicitly stated in an account of this sort. It was my impression that drinking did in fact appear to be well integrated in Rhodesian African life. I did not feel that urban drinking presented a serious social problem, although I do not mean that it was not a problem for some individuals. The municipal beer gardens appeared to provide a satisfactory solution to the problem of manufacturing, distributing and monitoring drinking in the townships. On a more personal note, I should add that I enjoy social drinking with friends and am inclined to think of drinking as one of life's pleasanter gifts. During my stay in Rhodesia I enjoyed many opportunities to participate in the renowned and genteel hospitality of White Rhodesians at "sundowner" time (or what Americans refer to as the cocktail hour). In the course of my lecture time I gathered the impression that alcohol plays an important part in the social life of Rhodesia's Whites, just as in the course of my more systematic inquiry I learned that alcohol, particularly beer, plays an important part in the social and ceremonial life of its Africans.

A Segregated Society: The Political Context

Perhaps the situation differs elsewhere on the African continent—although I rather doubt it—but it is certainly true that politics enters into virtually every topic of discussion and inquiry in contemporary Rhodesia. "As long as you don't get on the subject of politics, you'll find Rhodesians easy to talk to," I was often advised. Yet it was White Rhodesians themselves who inevitably sought an assessment from the observer in their midst, and their frequent "How do you like Rhodesia?" was an invitation for a sympathetic response to the real question that Rhodesians wanted to ask but had learned not to ask of outsiders: "Under the circumstances, don't you think we are doing pretty well—for ourselves, for Rhodesia, and for our Africans?" For an entirely different reason, political questions were equally conspicuous in conversations with Africans: such topics were never raised by them, never responded to if initiated by me.

The most recent turning point in Rhodesian history at the time of my arrival in 1970 was the Unilateral Declaration of Independence, referred to as "U.D.I.," which had occurred 5 years earlier when Rhodesia declared independence from the United Kingdom. U.D.I. signaled a dramatic departure from years of slow and uneven progress in the transformation from a colonial outpost to independent status, epitomized in the British goal of "unimpeded progress toward majority rule." Britain refused to acknowledge Prime Minister Ian Smith's new Government of Rhodesia and, with enthusiastic support from emerging nations in Black Africa, set in motion the imposition of international political sanctions aimed at isolating (or economically "strangling") the new country until it gave evidence of steering a more egalitarian course.

Economic sanctions may bring some political satisfaction as a manifestation of power, but they enjoy little reputation for actually accomplishing their purposes. Indeed, in the Rhodesian setting, sanctions were aptly described as being counterproductive. Efforts that might ordinarily have gone into expanding export industries were turned instead to manufacturing at home many products that had previously been imported, and the labor market most seriously affected was that of the unskilled urban African. A nervous and conservative government gained confidence as the economy struggled to survive and became increasingly conservative as it realized its own strength. Strong ties with neighboring South Africa grew stronger as the latter assumed responsibility for providing Rhodesia's economic lifeline. If Rhodesia of "Federation Days" in the late 1950s had seemed a model for Black and White partnership, Rhodesia of 1970 seemed in little hurry to have both partners share equally.

To most Rhodesians, the actuality of majority rule seemed remote, and even the possibility of achieving "parity" between the races—Africans and Europeans each assuming 50% of the political power—seemed years away. A period of political activism among urban Africans in the early 1960s which resulted in rioting, mass arrests and permanent internment for some African leaders had left certain lessons firmly imprinted among Rhodesia's countrymen. A prevailing skepticism that Africans were ready to handle their own affairs was substantially reinforced by that course of events and had become a rationale for proceeding cautiously. Avid supporters of the major political party, the Rhodesian Front, heard what they wanted to hear in policy pronouncements assuring "no forced integration." The efforts of liberal Whites interested in seeing Africans take a more active role in the government (and it should be pointed out that there were liberal Whites in Rhodesia, and that there were African members of the national parliament) met with apathy and suspicion. Older Africans who had grown up under colonial rule had recoiled at the violence and coercion which accomplished political activism in the townships and hoped such days would not recur. Younger Africans dared not voice their political opinions except in the most trusted company. Whether or not it was true, the belief that there were informers "everywhere" served to enforce political silence. The term "informer" need not conjure up an image of an army of full-time spies. A depressed economy and officials who will pay for useful information are sufficient ingredients to make everyone suspect everyone else.

The critical dimension in the segregation between Whites and Blacks in Rhodesia was not in where one was allowed to go but rather in where one was allowed to remain. Because of their witespread employment as domestics (with important distinctions between skilled and unskilled status), Africans were found virtually everywhere in Rhodesia—the best homes, the best clubs, the most prosperous offices. Their presence side-by-side with Whites was most often in subservient roles but was not invariably so. Africans could shop in many stores in the European areas. They used the same government offices and shared the same public transportation systems, although fare differentials in the latter imposed segregation along economic lines that tended to be racial lines as well. Africans not only served but also were served in many of the best hotels in the country. There was some socializing between Africans and Europeans (Africans being invited as guests in European homes or to formal occasions), although not as much as in the past according to observers. But when it came time to retire for the evening, then people returned to their proper areas. With the exception of Africans employed as domestics in suburban areas or as laborers on European-owned farms (situations where housing was provided as a condition of employment), Africans returned to quarters in township areas or remained in rural areas specifically designated for them (e.g., the communally held Tribal Trust Lands, or Purchase Areas where Africans were permitted to own farms in their own name). There were few settings where racial lines were relaxed—the interracial dormitories at the University of Rhodesia was one such place, and Whites and Blacks alike who hoped to see an eventual softening in the strict separatist climate watched the university "experiment" closely. Some critics argued that were it not for underlying economic considerations, the government would have turned the university into an all-White school, although others argued that its presence as a multiracial institution was one of the country's best answers to the constant international criticism leveled against its racial policies.

Integrated Drinking in a Segregated Society

It is my observation that drinking anywhere in the world tends to be an exclusive social activity engaged in by relatively closed groups of people who share some mutual elements in their lives. It was certainly true that drinking in Rhodesia was exclusive, ordinarily segregated along social-class lines as well as along prevailing "color" lines, and I hope I have not created the impression among readers unfamiliar with southern Africa that Rhodesia's Whites and Blacks drank in *socially* integrated settings.

The antithetical terms "integrated" and "segregated" help to dramatize the *cultural* functions of African drinking in Rhodesia in the context of the politically segregated society in which that drinking occurs. I would also call attention to the interplay between these two processes; that is, the political system that had produced a segregated society exerted important influences on the perceived functions of drinking. For example, some Africans searched for an ulterior motive to explain why Europeans who were in many ways so reluctant to share the advantages of their life style were so eager to provide the indigenous population with all the beer it could consume. Further, the effort which the European community expended, the rationale which it presented, and some of the benefits obtaining to both White and Black communities, contributed to a sufficient degree of "integration" between those communities to make segregation itself workable. That is, for better or for worse, the municipal beer industry provided an integrating link that contributed nobly to maintaining the status quo in the relationship between Africans and Europeans. As anthropologist Ralph Linton noted years ago (78, *p. 357*), one of the paradoxes of a high degree of integration among the elements of a culture is that it is balanced by a corresponding loss of the ability to alter the culture rapidly and with a minimum of discomfort to the society's members. Thus the very level of integration itself may—and must—be viewed for both its integrative and its disintegrative functions in the African and European communities. In rather diverse ways, African beer functioned culturally for the whole Rhodesian community in the same way as it functioned psychologically and physiologically for most individual drinkers: it facilitated rapport and prompted social solidarity; it dissipated some pent-up hostility and frustration; it enhanced gaiety and exuberance; and it contributed to accepting things as they were. These are the socially integrative functions of drinking as identified in cross-cultural study. In Bulawayo these integrative aspects had components "on both sides of Lobengula street."

The descriptive account proceeds now in four major sections. Part II, *Shared Behavior*, surveys the cultural milieu in which urban African drinking occurred, emphasizing the integration of beer drinking with other aspects of contemporary African life. Part III, *Professional Involvement with Urban African Beer Drinking*, examines urban drinking from the perspectives of a number of different formal roles in which attitudes and behaviors were related more to occupational status than to personal opinion. The nature and extent of this involvement illustrate how the operation of a municipal beer industry contributed to a degree of integration sufficient to maintain a formally segregated society. Part IV focuses on *Individual Adaptation to Urban Drinking* among township inhabitants. A final section, *Analysis,* reviews the major themes that seem to pervade the descriptive account and strikes a sort of cultural balance sheet to identify major integrative and disintegrative functions performed by urban African drinking for both White and Black Rhodesians.

Part II

Shared Behavior

2

Environmental Factors: Historical and Physical Aspects of Urban Beer Drinking

BUREAUCRACIES are not reticent about recording their achievements. The municipal administrations of Rhodesia's urban centers had dutifully recorded and frequently publicized their efforts to change and improve conditions for their urban African populations. The following account of the development of the municipal beer industry is derived from several sources provided by Bulawayo and by the capital city, Salisbury.

African Beer

A brief description of the process and product in brewing African beer appeared in an attractive brochure, *Rufaro, The Story of African Beer in Salisbury* (110, 111), prepared after the opening of the new Rufaro Brewery in the capital city in 1964:

> To a certain extent the basic principles of brewing African Beer are similar to those of the conventional Brewing industry: mashing, conversion, cooking, fermentation and filtration are vital processes common to both. There are, however, substantial differences and African Beer is quite different in taste and appearance from conventional European-type beer or ale.
>
> Generally speaking, the brewing of African Beer is very much simpler, and the period of time is much shorter. No hops are used. The taste is slightly sour, rather than bitter, due to the lactic acid used in the brewing process. The whole process takes about 60 hours, as against eight to ten weeks for a lager beer. African Beer is not stored prior to consumption, and the need for expensive storage cellars is accordingly obviated. Fermentation of the beer continues right up to the time of consumption, and the final product contains about 6% of solids in the form of suspended cereals. This is how the customers traditionally want their beer.[1]

[1]Liquor Undertaking Department (110, *pp.* 4–5); paraphrased.

A more detailed account of the brewing procedures followed in Bulawayo and of the raw materials used is provided in Appendix A. Laboratory specifications of the final product are also included in that account. Varying opinions regarding the taste and appearance of the beer are provided in the pages that follow.

History of the Brewing Enterprise in the Municipalities

In their earlier days, Bulawayo and Salisbury followed a rather parallel course of development as Rhodesia's two major cities. Whatever either city adopted or pioneered, the other usually incorporated soon after. In the case of developing a municipal brewing industry to produce African beer, Salisbury took the first steps in 1908. Bulawayo followed 5 years later. More recently, and especially since the Rhodesian declaration of independence in 1965, Salisbury's growth and favored position as capital had become obvious to Bulawayans, and the latter had assigned the nickname Bamba Zonke, "grab everything," to their seemingly more favored sister. Bulawayans pointed with pride to the new and attractive National Museum of Rhodesia located in their midst and pondered, "I don't know how we happened to get this—it must have been a mistake."

Contemporary rivalry between the two cities did not appear to be divisive but it did result in a constant tendency for drawing comparisons between them. In the face of Salisbury's obvious lead (its European population was almost double Bulawayo's 50,000, its African population 1½ times greater than Bulawayo's 190,000), Bulawayans were quick to call attention to any facet of municipal life in which they equaled or excelled their more favored twin. Their success in developing a municipal brewing industry equal to Salisbury's provided them with considerable satisfaction in this regard.

Both communities started their beer "industries" on a very modest scale by employing African women already recognized for their brewing skills to produce beer in sufficient quantity to be marketed commercially. The combination of rapid growth of the populations in the African townships and the initial success of the beer-making enterprise encouraged the development of more efficient brewing techniques and facilities and the expansion of retail outlets as the areas encompassed by the townships continued to expand.

The Bulawayo Municipal Liquor Undertaking provided the following brief account of its progress (16, *p. 1*):

"The Bulawayo Municipality built its first African Beer Brewery and Beer Hall as a combined brewing and selling unit in 1913. This served satisfactorily until rapid growth in urban population in the late 40s made it necessary to establish a separate brewery and to start on an expansion program that is still not complete. . . . Large scale adaptations and improvements are being introduced at the Brewery with a view to doubling our capacity over the next few years."

Bulawayo produced some 7000 gallons of beer in its first year and over 6 times that amount the second year.[2] Average annual production rose steadily and showed a dramatic increase with the opening of the new Steeldale Brewery in 1951, when sales jumped from 890,000 gallons to 1.5 million in 1 year. In the fiscal year ending in 1970, Bulawayo produced 14.5 million gallons of beer, with sales slightly in excess of $3 million (Rhodesian) for draught beer alone, as in Table 1. Gross income from the Liquor Undertaking was further augumented by other beverage sales which illustrate not only the magnitude of the industry but also reflect beverage preferences of the African clientele at the various municipally operated beer gardens and bottle stores.

TABLE 1.—*Income from Municipal Beverage Sales in 1970*

Draught beer	R$3,033,493
Packaged beer	237,740
European beer	498,755
Spirits	72,949
Wines	149
Soft drinks	4,557
Total	*3,847,663**

*Approximately U.S. $5.4 million.

The Contemporary Scene: Facilities

Outlets for beer in the African Townships of Bulawayo included 1 beer garden complex (known as Main Beer Hall because it was originally both the brewery and the only beer outlet), 30 beer gardens and 4 bar-lounges. Bottled and packaged beer was also available for purchase to take out at numerous outlets, including 11 bottle stores, all 4 bar-lounges, and most beer gardens. In addition, the Steeldale Brewery (more commonly known as the African Brewery

[2]All volume measures are in Imperial measures: 1 Imperial gallon = 1.2 U.S. gallons = 4.5 liters.

or simply The Brewery) had 25 regular nonmunicipal customers to whom it sold and delivered draught beer in bulk including Army, Police, Prison and Local Government Authorities, several rural African Councils and a number of large employers of African labor who made beer available for purchase by their employees. The capital investment as of June 1968 represented about R$2,024,600 (approximately U.S. $2,834,000), with slightly more than half the sum invested in brewery and distributing equipment, the balance invested in the beer gardens and other sales outlets (16, *pp.* 2–3).

Recognizing "rapid technological developments" in the field of brewing, Bulawayo had minimized the amount spent on buildings at its brewery in order to allow flexibility in making large-scale adaptations and improvements in brewing procedures. An ambitious list of actual and projected improvements included doubling the boiler capacity and providing all boilers with automatic stokers, improving the malting and milling plant, increasing grain storage capacity (to minimize the effect of fluctuations in availability and cost) and handling facilities, developing a new bottling and packaging plant, and adding new cooling towers.

For distributing its bulk beer throughout the townships and to its other clients the brewery had a fleet of 13 tank trucks with a capacity for transporting in excess of 15,000 gallons of draught beer at one time. Total beer garden storage capacity was 60,000 gallons of bulk beer.

With the one important exception that beer was stored in huge tanks located overhead, thus allowing for gravity-flow dispensing rather than pumping, the distribution of African beer from processing plant to consumer was similar to procedures at a type of gasoline "filling station" familiar throughout much of the world. Beer was delivered at least once a day to each beer garden and was pumped from the delivery trucks directly up into overhead storage tanks in covered lofts. The serveries were located at ground level under these storage lofts.

Customers at the beer gardens queued up outside the servery area to purchase their beer. Ordinarily they made their own selection of a half-gallon or gallon plastic container from a rack located nearby and adjacent to mug-washing facilities. The customer handed an attendant his empty "beer mug" and his money. The attendant made change as necessary and then filled the mug with a quick 180° rotation of the Buck meter, elevating a measured amount of beer into the upper position so that its contents emptied through an open valve into the waiting mug, while the unit at the opposite end of the arm filled automatically by gravity flow in preparation for the next sale (this is illustrated in Plate 4). The effect was like rotating a beer-filled and rapidly emptying hour glass. Each dispensing unit consisted of two Buck units fed through a common pipe. By custom, the Buck meter on the right-hand side delivered a 2-shilling measure of beer (1 gallon), and the meter on the left-hand side delivered a 1-shilling measure (or, in less affluent areas, a half-shilling's worth).[3] A sealed tumbler-type counter fitted onto the arm of each Buck meter recorded the number of units of beer dispensed.

The number of dispensing units at any one beer garden and the number of attendants on duty at any given time was a function of the size of the beer hall, the hour of day, and the day of the week. Beer gardens that operated from midmorning were not particularly busy during the first few hours on weekdays. All gardens were busier in the late afternoons after normal working hours. Saturday and Sunday were the big days for social drinking. At the more central locations, the volume of beer sold was incredibly large; the busiest outlet in all the townships, one of the serveries in the Main Beer Hall, dispensed up to 6000 gallons of beer a day, and the combined serveries at the Main Beer Hall had reached a record high day selling 18,000 gallons. The number of persons in and about the beer gardens could also reach incredible proportions. On a hot and busy Saturday on a payday weekend in 1970, the superintendent of the Main Beer Hall organized a head count to ascertain the number of persons passing through any of its 5 gates. Recognizing that some persons were inevitably counted more than once and that persons merely walking through the area could not be excluded from the count, there was still some surprise in the total reported tally of 52,000 persons in a 12½-hour period from 9 AM to 9:30 PM.

Inside the beer gardens, customers sat about in a large enclosure that constituted the open-air courtyard of the garden (Plates 5, 6, 7). There were several alternatives for seating arrangements and gathering places. In hot weather the most critical factor was to find a place in the shade; occasionally protection from wind or rain was important, but rainy or cold weather usually kept people at home as predictably as hot weather found them flocking to the beer gardens.

[3]The shilling was worth about 14 U.S. cents.

Every beer garden had a series of covered sheds or "shelters" dotting the periphery of the garden and usually consisting of one or more walls and a roof. These shelters were popular places for gathering. Some gardens had especially inviting features, such as seats built around existing rock outcroppings or under shade trees, or were adjacent to park areas with inviting lawns; those who came late had to be content with some arrangement of the available benches in the open areas. Many gardens were surrounded by high stone walls. These were designed to provide privacy and, to a limited extent, to serve as sun and wind breaks rather than to keep people out; the entrances to most beer gardens were never closed and many of them served as popular (and interesting) short-cuts through township neighborhoods.

There was considerable variation among beer gardens due to differences in their location and terrain, in the nature and extent of trees and turf, and in the architectural spirit predominating at the time each was constructed. In general they were stark and utilitarian, dominated by the bare earth under their milling hordes (there was a saying heard in some circles among White Rhodesians that "grass will not grow where Kaffir feet tread"), and yet here and there one caught a glimpse of some consciously placed accoutrement like a tiled or painted mural, an extra wall, a thoughtfully designed seating arrangement of new benches and tables. Each alternative place for gathering had its special adherents. The sheds were particularly coveted as the sanctuary of groups of regularly gathering associates generally recognized as having territorial rights.[4] Especially on weekends, some people made a point of arriving at their local beer garden early in order to hold a coveted or customary meeting place, and well-established groups of drinkers ensured that their regular sheds were reserved for them by designating persons responsible for an early morning "booking" of space with such acceptable markers as newspapers, shoes or other articles of clothing, and even inexpensive decorations of home-made bunting tied to the rafters for the weekly rendezvous. Such coveting of space was not without rivalry among groups and complaints to managers, but neither were officials unaware that the groups brought in many patrons, bought large quantities of beer, and were, as large groups of beer drinkers go, usually very orderly.

[4]Cf. the Austrian *Stammtisch* tacitly reserved for regular customers; see also Sommer (124, *ch. 8*).

The hours when beer could be purchased varied slightly from one beer garden to the next. Staggered hours between adjacent gardens made it possible to operate most outlets with one 8-hour shift of 8 to 11 employees and yet keep service conveniently available somewhere in each neighborhood throughout the day and into the early evening. At the Main Beer Hall, for example, each of the 3 serveries was operated as a separate unit with its own staff and its own hours. The earliest opening unit began serving at 11 AM, the next at 12:30 PM, the last at 2 PM. Thus service was available for 10 hours, although each servery operated a shift of about 7½ hours on weekdays and 8½ on weekends. A comparable pattern of opening and closing prevailed throughout the townships. The larger beer gardens opened around noontime and closed about dusk, but like the theoretical 24-hour pub crawl said to be possible in London, someone who knew his way around the townships could purchase African beer on sale from 9:30 in the morning until 10:30 at night every day of the year. Opponents of the liquor undertaking cited this "easy availability" in criticism, although their arguments did not usually include two corollaries; first, that in order to avail oneself fully of the opportunity for such continual drinking, one would need money and private transportation, both always in short supply, and second, that drinks could also be purchased virtually round-the-clock at private parties in "shebeens," one of several alternatives to going to the beer gardens.

Beer-garden drinking was an unhurried affair. Although table service was available in the bar-lounges and in some specially designated area in the larger beer gardens for the purchase of European liquor and packaged African beer, draught beer was the major product sold and it was provided only on a self-service basis. It was possible to spend a day in the beer garden just visiting, watching, and listening to the crowd (and, sometimes, listening to recorded or live music varying from electronic "rock" bands to tribal drumming) without making a purchase or feeling any obligation to do so. The vast majority of beer-garden patrons, however, were there to drink their portion of the millions of gallons consumed annually. The average annual per capita consumption of African beer among urban male drinkers was estimated at about 185 gallons (86, *p.* 5), a volume that concerned officials not only for its magnitude but also for the incredulity of attempting to arrive at such an estimate.

The custom of passing the heavy-duty plastic beer mugs from person to person among a group of drinkers did not appear particularly hygienic (although it is not necessarily less so than in other collective uses of a single source of stimulant, whether it be sharing a peace pipe, a "joint," or a chalice) but throughout the entire brewing and distribution procedure there was an evident effort to maintain standards of cleanliness. From the beginning of the brewing process to the moment when the beer was dispensed, it moved through a labyrinth of carefully scoured pipes and tanks. Brass fittings and exteriors on the dispensing machines usually sparkled in conspicuous proclamation of cleanliness, and the serveries were hosed and scrubbed daily. Care was also given to washing the beer mugs, although the results were not quite so impressive. Employees assigned to the endless task of mug collecting made continual rounds in the beer halls to gather empties and return them to stainless-steel washing facilities where they were hand washed, rinsed, and placed on open drying racks in stacking areas convenient for the next purchaser. Fastidious customers chose mugs with care from the sometimes motley assortment of empties on the racks and voiced complaints over the perfunctory manner with which mugs were washed. Patrons and authorities alike expressed exasperation with the condition of older mugs that so frequently showed evidence of vandalism (knife slashes, crushing and splitting, melting with a flame) as well as ordinary wear and tear that led to a perennial problem of mug losses at a rate calculated at 14% a year. One of the duties of the special municipal police assigned to the beer gardens was to keep an eye out for destruction of property, a problem to which the plastic mugs and moveable benches were easily the most susceptible.

The Contemporary Scene: Policies and Trends

Official enthusiasm for progress and potential in developing the Liquor Undertaking was expressed in a major report prepared in 1957 that dealt with African administration in Bulawayo. The report emphasized many aspects of the beer gardens (15, *Appendix C*):

> "The pastime which caters for more people than any other, and which most prevents trouble is the beer drink. The Council's policy of beer gardens has enabled greater progress to be made in this direction than any other Local Authority in South Africa. Its wisdom has been so clearly demonstrated. There is no need to recapitulate its advantages.
>
> "This policy should be continued and additional beer gardens opened in all new areas, and wherever congestion in existing gardens warrants.

"The recent improvement in brewing has led to greatly increased beer sales and patronage of the Gardens. Both would be still further increased by improving the Gardens in various ways. Some of these, such as additional lavatories, wind breaks, and shelters have already been introduced [at some sites and] should be extended immediately to others.

"Hitherto there has been a tendency to regard these centres simply as beer drinking places. But they are much more than this—they are real recreation centres where people can relax, gossip, have small parties, sing and dance. This aspect should be greatly encouraged. Removable screens should be provided for organized private parties, and open air dance floors constructed which can either be hired out (as at the Central Beer Hall), or where our own dance bands or hired dance bands can provide free entertainment.

"More attention should be paid to the appearance of these centres to raise their tone and encourage the attendance of many self-respecting people who at present avoid them."

The same report also noted what was probably one of the most important elements in maintaining the popularity of African beer: "Since 1913, beer has been sold at a standard price of 2/- per gallon" (15, *Appendix L*). Even with the continually rising prices of the 1960s the beer price was maintained at 2 shillings. When Rhodesia converted to decimal currency in 1970 the price remained 20¢. Officials had finally acquiesced to the pressure of rising costs, however, and they had decreased the volume of beer dispensed by 10% and planned another slight decrease to coincide with the national conversion to metrication in 1971. This created some confusion in accounting and reporting, since "gallons" recorded through the meters and gallons delivered were no longer comparable. Dollar sales provide the most accurate basis for year-to-year comparisons. Following the 10% volume reduction, officials were inclined to note that the *price* of African beer had not changed since 1913.

Concomitant with stressing the social aspects of the beer gardens and the successful attempt to hold the price of beer constant, officials concerned with Bulawayo's beer industry prided themselves on three other facets which reflected their efforts on behalf of the Liquor Undertaking: constant improvement of the quality and availability of services for retailing beer; the quality of the beer itself, particularly including its nutritive value; and the welfare services and other amenities made possible by beer profits.

Attention to the services for providing beer were directed particularly to making the beer gardens and bar-lounges attractive places and for increasing the number of beer gardens in proportion to township expansion. The report cited above summarized the department's policy (15, *p. 39*):

"Efforts are being made to get these Beer Gardens to be recreation centres rather than drinking dens. They are in the open, with shelters against wind and rain. Trees are being planted for shade and appearance, and dry [i.e., nonalcoholic] canteens, run by African traders, supply soft drinks, snacks, and cigarettes, so that people can meet their friends and have such refreshment as they desire in pleasant surroundings. They have been immensely popular, and more are needed to stop the pleasant atmosphere being spoiled, on which depends their success as recreation centres."

At the time of the 1957 report, the municipality was operating 15 on-sale premises in the townships, and 3 new beer gardens were under construction. By 1970 there were 37 such outlets plus 11 bottle stores for off-sale purchases. Development was proceeding along 2 lines. One was the provision of modern lounge-type facilities, catering to a sophisticated African clientele interested in table service, live music and Western-style dancing in an atmosphere comparable to a European-type cocktail lounge or night club. African beer was not available in draught form in the bar lounges, although it could be purchased in packaged form along with bottled European-type beer and spirits. One of the bar-lounges had become somewhat of a showplace for demonstrating to outside visitors not only the increasing progress and sophistication evident in the urban native population but also the administration's accommodation and sensitivity to such change.

The other focus of improvement in service was the existing beer gardens, for although the officials in Bulawayo recognized the need for operating some facilities catering to a growing class of African élites (67) and probably recognized the desirability of having such showplaces for international observers as well, they were firmly convinced that the majority of the township population was better served through the beer gardens. In a report prepared for the City Council, the Liquor Officer specifically recommended "the provision of sufficient outlets to cater for the needs of small local communities" and "the improvement of beer gardens to meet the growing sophistication of the average working class African." In conjunction with his recommendation for concentrating primarily on the beer gardens, he added, "I regard this as more important than the provision of a greater number of high class lounge-type outlets" (57, *p. 11*), and his report, based on a series of brief visits to African breweries in Rhodesia and South Africa, contained numerous references to and comparisons with facilities in Bulawayo's beer gardens for providing live and taped music, improved décor and atmosphere and African-operated food and canteen services.[5]

Administrative concern for the quality of African beer dates back to the first venture into commercial production, although the nature of that concern has gone through several stages, each in part a reflection of changing moods and "growing sophistication" on the part of urban townsmen. Earlier, attention was drawn to maintaining a consistent quality in the beer produced and, at times, with producing sufficient quantity for the established and expanding market. As the magnitude of the enterprise had grown, a new problem known as "adulteration" had arisen. Adulteration refers to the practice of adding water to beer before it is sold and pocketing the profit from sales exceeding the volume that must be accounted for. A considerable portion of the efforts to maintain "security" in the beer gardens was actually a concern for protecting the beer from dilution by the beer-garden employees. Accusations that the beer had been diluted were still aired occasionally by patrons, although improved security plus wires and lead seals on valves and tank covers (and a recommendation that only tanks with bottom-opening manholes should be purchased in the future) had helped alleviate the problem. "We had to let 12 people go just yesterday for this sort of thing," the Liquor Officer explained on the first day we discussed the problem of beer adulteration.

Among the beer drinkers of Bulawayo, as among beer drinkers throughout the world, some batches of municipal beer never quite "tasted" right. Two factors had kept this problem to a minimum, although since the ultimate test of beer is how it tastes to the people who drink it, the problem can never be completely eliminated. One was the constant attention to brewing technology. The other was the feeling among Africans that their complaints regarding beer quality

[5]Although the Liquor Officer's job, as he described it, was "to distribute and sell beer," a content analysis of his report prepared for the City Council showed that almost one-fifth (19%) of his entries dealt with the attractiveness of the on- and off-sale premises and the provision of amenities like music, food and service. Comments related to the esthetics of the beer gardens comprised one of the single most frequently mentioned categories in the report (7%), exceeded only by mention of mechanical facilities for security, storage and handling of draught beer within the beer gardens (9%).

were heard and investigated. As a matter of fact, some Africans noted a contrast between the attention given their comments about the quality of the beer with their requests concerning the quality of other aspects of township life. An informant noted, "Sometimes the beer is not made right, and people complain. Then the Superintendent takes great pains to see that it is made right. But if you say, 'Our roads are no good. Something should be done,' they just say, 'I'll see about it.'" As administrators noted, it is considerably easier to "do something" about a batch of beer than it is to put in a new road. Nevertheless, attention to the production of a good beer of consistent quality remained a stated objective and, with constantly expanding volume and seasonal variations in ingredients, a constant challenge.

The most recent concern for beer quality had been with the "image" of African beer. As already noted, officials in Bulawayo were committed to expanding their African beer industry along with catering to the gradual but apparently inevitable increase in the percentage of the total market given over to European-type beers. Until 1957 Rhodesia's Black population had been prohibited by law from purchasing European-type beers and other spirits. Since that time, the popularity of European beer had increased continually, and the municipalities in Rhodesia and in South Africa which produced and sold African beer looked with varying degrees of enthusiasm on the problem of marketing more potent beverages which they could offer for sale but did not brew. All Bulawayo's beer gardens had at least a modest facility for selling bottled European beer, which was sold at the prevailing local rate of 18 to 20¢ per 12⅔-ounce bottle (compared to 20¢ for a gallon of African beer).[6]

Official policy in Bulawayo was that African beer should continue to hold the major part of the market, and to this end local administrators constantly sought ways to maintain and improve the image of "their" beer in the face of trends elsewhere to facilitate and even to foster the transition to European beer. After a formal inspection of facilities in Salisbury, the administrative officer for Bulawayo's Liquor Undertaking reported several factors operating there which he felt to be instrumental in accounting for Salisbury's

[6]In order to ensure the return of empty bottles, a 4¢ deposit was charged on every bottle of European beer—as well as on glasses from which to drink it at some outlets. Africans reporting on their expenditures for European beer typically included the deposit as part of the purchase price. Thus a man who drank one bottle a day would report spending 22¢, a man who drank two bottles, 44¢.

relatively smaller proportion of African and higher proportion of European beer sales compared with Bulawayo: the sparser distribution of beer gardens throughout the townships, the predominant use of half-gallon mugs for dispensing African beer, and the greater emphasis on cocktail bars and lounges designed to appeal to élite patrons (58, *p. 4*). The report noted: "Whilst Salisbury probably has a larger sophisticated element amongst its patrons than does Bulawayo, it has a much larger unsophisticated population which is inadequately catered for and which is, therefore, to a degree induced to turn to European beer."

During the Liquor Officer's subsequent visit to several African breweries and beer gardens in South Africa, he found the trend observed in Salisbury even more pronounced. Inadequate facilities for keeping up with the demand for African beer, and poor facilities for its on-sale consumption in most municipalities he visited, contrasted sharply with the reasonably good facilities for handling both the on-sale of European beer and spirits and with the "high image" given these products. He found the whole native beer industry in South Africa overshadowed by the conviction that the brewing of African beer (known locally as Bantu beer) as a commercial enterprise had passed its prime: "A number of the officials I met were convinced that the dictum put forth by South African breweries that European liquor would wipe out African beer within a decade would inevitably come to pass" (57, *p. 11*).

The most-touted virtue of African beer on the part of those who supervised its manufacture and sale was its nutritive value. The European staff in the Housing and Amenities Department extolled the high food content of the beer as contrasted with its low alcohol content. A comment volunteered by one official illustrated his enthusiasm: "Some of these fellows who work as laborers take only African beer for their lunch. That's their entire meal! But the beer is so nutritious that it provides their food as well as their drink."

A comment in an attractive illustrated brochure describing Salisbury's Rufaro beer provides a similar example. This excerpt was taken from a Foreword prepared by His Worship the Mayor, an ex-officio member of the municipality's Liquor Undertaking Committee (110, *p. 1*): "Rufaro beer, besides being a beverage, is also a food —and some of these pictures will give ample evidence of its valuable contribution towards the health of our indigenous population."

Newspaper advertising for an African beer produced by private

Rhodesian industry primarily for sale in packaged form in rural areas similarly emphasized the well-being of the drinker: "Every glass of Chibuku is as good as the last one, for Chibuku never varies, and you can drink as much as you like because Chibuku helps you to wake up fresh and healthy next morning."

With equal vigor, municipal administrators argued the merits of their beer industry in terms of the underwriting of housing, health and welfare services made possible by the beer sales. A brief report issued after the close of the fiscal year 1968 summarized the extent of the profits and the major uses to which they were put (16, *p. 4*):

> "These surpluses which now run at ± £500,000 p.a. [U.S. $1,400,000 per annum] are used in the day to day provision of health, recreation and welfare services, and on capital provision of clinics, schools, libraries and recreation facilities and housing. This last is now assuming greater importance following the Government's decision to allow up to one third of the surplus to be used for this purpose."

The contribution to Welfare and Health Services for 1967–68 was reported at £314,456 (about U.S. $880,000). These monies contributed to a vast network of services, including the provision of maternity clinics, the salaries of welfare case-workers, youth club programs, and subsidies for a "Home Industries" enterprise which provided regular employment for members of exceptionally needy families. A detailed account of these facilities has been compiled in a major report, "Welfare Services in an Urban African Area" (35). The crucial aspect in terms of the beer-drinking industry was that, as previously noted, the municipality earned a huge profit on the enterprise but turned that profit back into the townships to provide housing and amenities both in services and in funds for capital investment. Recent capital works (1965–68) had included sports facilities (dressing rooms at a central stadium, a new swimming pool), improvements in arts and crafts facilities and centers, two youth centers, two libraries, additional park facilities, a medical clinic and a new secondary school.

As soon as beer profits became big enough to notice, they became a matter of controversy. The allocation of profits was one of the problems associated with the beer industry that had steadfastly maintained its omnipresence but frequently changed its form through the years. An argument put forth by an early Government Commission maintained that beer monies should be spent only on the "betterment of Native conditions in the Location." The Bulawayo City

Council, which had always followed such a policy in fact, argued against the Commission's findings in principle on the basis that such a policy would saddle the European sector of the community with forever having to "bear the burden of Town Management, while the revenue from purely Native sources shall be expended solely for the benefit of Native conditions in the Location."[7] The City Council pointed to the fact that National Government derived certain of its revenue from purely Native sources but did not expend them on purely Native purposes. "This exchange of opinion," it was reported, "raised matters of fundamental principle which have never really been considered or decided upon, and also created a bone of contention over which the Government and Municipality are still wrangling" (15, *p.* 37).

Formal commissions of inquiry, outspoken critics and their own "tender consciences" had served to keep members of the City Council and officials of the African Administration ever mindful of problems created by life in the townships and of disruptive aspects directly traceable to or felt to be associated with the increasing beer traffic. The 1957 report included reference to a familiar and recurring theme in the criticism of the Municipality's beer industry: "the 'beer hall traffic' pointed to the steady demoralization of the people in spite of the activity of the churches, and the beer habit so established was largely responsible for the poverty and domestic unhappiness found in the Location." The report made no attempt to deny such problems; rather it emphasized official concern for them. "This is an extremely important aspect of the 'beer traffic,' one that is worrying us today with the steadily rising consumption of beer and the recent introduction of European beer and wines" (15, *p.* 36). Elsewhere in the same report (*pp.* 55–56) the administration reviewed its hopes and efforts to recognize and ameliorate the undesirable aspects of township life through developing a vigorous program in Welfare and Recreation that would

> "cheer up the African townships—to prevent, or arrest, the disintegrative forces of town life, which are accentuated by the abrupt transition from tribal, rural existence; and to counter the drabness of town surroundings and the dreariness of industrial experience; to offer a positive release to frustration and bewilderment occasioned by these factors and exacerbated

[7]Jackson Commission Report. Quoted in a Bulawayo City Council report (15, *p.* 37).

by racialism and daily humiliation; [and] to provide decent, gay, attractive, constructive recreation as an alternative to beer swilling and fornication."

Another problem which can be included in an examination of the official position of the Municipality, one sufficiently stressed in word to suggest some uneasiness regarding the status quo, yet itself another problem without easy solution, is the issue of competition. Wherever they were established, the municipal beer industries represented virtually a local monopoly. The fact of the monopoly was a convenience, but the accusation of maintaining a monopoly was not necessarily a comfortable one, particularly when it could be cited by the African population as another example of the total domination of their lives by the European community. Competition was not entirely absent, however. Competition was inevitable given the interest of the brewers of European-type beers in capturing more of the urban African market. The well-known Chibuku beer, produced by private industry for sale in rural areas outside the protected urban markets, was at least allowed to advertise locally. Operating its own chain of 37 on-sales outlets, the Bulawayo Municipality boasted magnanimously (16, *p.* 3): "Nor are we afraid of private competition in this field. We have built and leased to Mr. Vera, the Happy Valley Hotel, and we have leased land to Mr. Zwambila where he has built the Marisha Cocktail Bar, which is one of the most attractive of its kind in Rhodesia."

At the same time that competition was "welcomed," the Bulawayo Municipality had become more aggressive in its own sales efforts. In the townships, it sought to improve services to make its facilities increasingly attractive. One observer noted that such efforts were not spared at beer gardens in the vicinity of the two privately run facilities. The municipality also operated under the economically protective aegis of an old law which prevented privately owned liquor outlets from selling the least expensive and locally most popular alcoholic beverage, municipally produced African beer. As the distribution of commercially produced African beer continued to make inroads into the tribal trust lands, the municipality paid more attention to making its own beer available in outlying rural areas within a radius of many miles from Bulawayo. "After all, we've got to protect our territory," liquor officials explained. Sales representatives met frequently with (and provided samples of beer for) councils of rural chiefs who already did—or might be induced to—purchase beer for retailing to their local constituents through community-operated outlets.

Salisbury had recently adopted a different strategy toward local competition, one more bureaucratic but less ambivalent. It publicly proclaimed its Liquor Undertaking Department, which operated as a separate department of the municipality rather than as a section of a comprehensive African Administration, a "benevolent monopoly" and pleaded for public sympathy and Government protection for this position. One alderman, a member of the City's Liquor Undertaking Committee for 1969–70, spelled out this position in a Foreword to one of that municipality's public relations brochures (111, *p. 1*):

> "It is important that the central Government in its wisdom recognizes the value of this source of revenue for there is always an avalanche of requests for disbursements from the beer funds. I am hopeful, therefore, that Government will do all it can to help, for I believe that the interests of this benevolent monopoly should take precedence over those of private enterprise which has openings in countless other fields."

Alternative Forms of Drinking for Urban Africans

An urban African who wanted to enjoy a beer in Bulawayo or to enjoy an atmosphere sparked by the sociability which typically (or at least ideally) pervades most of the world's social drinking had alternatives other than going to the local beer gardens or the more formal bar-lounges. Although this account is primarily concerned with drinking related to the beer gardens, the alternatives provide an essential complement to the total picture of urban drinking. Five such alternatives are described here: home brewing, off-sales purchase, drinking in the tribal trust lands, going to a shebeen, and drinking in the European areas.

Beer brewing is an age-old custom among the native people of southern Africa (Plate 1). Almost every account of African life from the earliest reports to the present day includes mention of the occasions for and prevalence of beer drinking. The beer is variously referred to as African beer, native beer, Bantu beer, or kaffir beer, the term kaffir, in the sense of "heathen," being the ignominious but nonetheless widely used term by which some working-class Whites refer to Africans.[8] In contemporary times brewing was not neces-

[8]The word is Arabic in origin, originally applied in a derogatory sense to any natives unconverted to Islam. As one early writer explained it, "The word *Kafir*,"

sarily an illicit activity; legality depended on where the brewing took place. In spite of the law, one heard reports of beer being brewed virtually anyplace where Africans resided. The basic process by which traditional African beer is made is described in Appendix A, although there are numerous variations in procedures and ingredients.

The manufacture and sale of intoxicating liquor for Africans had, as one might expect, "been the subject of legislation from the earliest days of the colony" (15, *p. 36*). Regulations established in 1895 prohibited the consumption of any alcoholic beverage in township locations—specifically including African beer. Those regulations were subsequently amended to exempt "Native Beer," but until the municipality established its African Brewery in 1913, home brewing and illicit drinking in the urban areas had been rampant. The art of brewing had not died out in the interim. Brewing had never been categorically outlawed throughout rural areas, and home brewing had always continued on some scale in the townships in spite of its illegality.

Certain periods in the history of the municipality saw a noticeable increase in illicit brewing in town or an increase in the amount of beer imported from outlying areas. Apparently one such period coincided with community reaction to the too-limited facilities for meeting the increased interest of women in obtaining beer in the late 1920s. The failure of the beer industry to keep up with the rapid urban African population growth of the war years (1939–45) evoked another similar era. In the latter case illicit brewing and drinking were reported to have reached such magnitude that great drinking parties with more than a thousand participants were not uncommon, and a special Police Liquor Detachment "was reduced to travelling in armoured vehicles and had sometimes to use tear gas to break up illicit gatherings" (15, *p. 38; 43, pp. 245-246*). A more recent recurrence of illicit brewing in 1953–54 was traced in part to a dissatisfaction with beer quality that was coupled with a degree of social and economic unrest not so easily remedied; subsequent improvement in brewery and beer-garden facilities, coupled with

used in its broadest sense, includes "all the dark-skinned tribes of South Africa. . . . The word does but mean *an infidel*, and has been frequently confined to a few tribes living south of Natal. But the title is often used in South Africa as a synonym for the word *native*" (66, *p. v*). The author also noted some variant spellings: Kaffir, Kafir, Caffre, Cafir, Kaffer, Caffir (*p. ix*).

renewed efforts to enforce existing regulations, led to a dramatic decline in convictions. But the very nature of African tradition suggests that there will continue to be brewing wherever there are Africans. Only when such activity reaches certain proportions does it arouse official concern.

Because brewing was not illegal in the rural areas, a return to one's home district in the tribal trust areas provided another alternative for engaging in a beer drink. Although some urban Africans confined their drinking to their visits "home," it is unlikely that they made such trips expressly to obtain beer. To the extent that individuals thought about beer in connection with travel between city and country, it was more likely related to purchasing packaged beer in town for subsequent resale at a profit in the rural areas. But there were countless numbers who did return to the country as often as possible and thus maintained their places in traditional life, a life in which beer had always played an important part.

There were Ndebele men living in Bulawayo who actually lived their lives as weekly "commuters," completing a work week in the city but returning each weekend to the rural areas where they maintained their families, their homesteads, their cattle and goats, and their agrarian traditions. One old man who had spent a career as a government clerk but had recently relocated in a rural area explained to me, "We *work* in the city, but we *live* in the country." The possibility of returning home on so frequent a basis was limited to the Ndebele. Bulawayo was located "in the heart of Matabeleland"[9] and distances were manageable via private transportation, the Rhodesia Railroad, and, especially, the ubiquitous African buses. Africans from the far corners of Matabeleland and those from more distant regions could not avail themselves of such frequent visits, but in my formal interviewing and informal conversations I met few Africans who considered Bulawayo to be their "real" home. As will be discussed in a later section, the economic and social necessities which draw so many men to the cities in search of employment and require them to remain apart from their families provided a sense of disquietude which many chose to dispel through beer drinking.

[9] I have followed the current usage in English of applying the term Ndebele to both the language and the people who speak it. The people are also known as Matabele. The terms derive from their language, *isiNdebele*, and their name for themselves, *amaNdebele*. Similarly I have employed the term Shona to represent the people and language of the Mashona tribe.

Home brewing was not prerequisite to home drinking. Bottled and packaged beers were easily available within the townships. Some patrons preferred to drink at home, purchasing their beer at the bottle stores or perhaps sending someone, often a child, to a nearby beer garden to purchase draught beer. Private house-parties were popular, particularly on weekends, the host supplying the beer for his friends. Semiprivate parties were also popular—parties at which everyone contributed toward the cost, or at which invited friends drank as guests of the host but uninvited guests willing to pay for the food and beverage were also admitted. In addition, the municipality at times made beer available at community-wide activities, a portion of the proceeds being retained by a sponsoring organization for carrying out community projects.

In addition to its draught-beer industry, the municipality marketed its beer in packaged form in three varieties, a pasteurized bottle beer, and beer in waxed cartons ("Conipaks") marketed under two different labels (Plate 9). Each packaged line had its own adherents.

The development of the bottle and package stores, particularly with the increasing popularity of European beers, caused some shifts in marketing practices. In Salisbury, bottle stores were usually located in complexes adjacent to beer gardens (Plate 10). In an effort to reduce the pressure on overcrowded beer garden facilities, the package stores handled a competitively priced and purportedly "stronger" beer available only in package form with the intent of fostering a new market among customers who would prefer to drink the stronger beer and would drink it at home. In fact, however, some patrons brought the stronger beer into the beer gardens and mixed it with the less potent draught beer, arguing that by doing so they had a stronger beer at lower cost.[10] Bulawayo patrons sometimes "laced" their draught beer by adding the purportedly stronger packaged beer to it. Another almost inevitable result of selling European beer at both on-sale and off-sale prices was that patrons created impromptu beer gardens in or near the bottle stores or purchased beer at the stores and carried it to the beer gardens to drink, practices easy to legislate against but not easy to stop.

Shebeens provided an interesting variation in urban drinking throughout southern Africa. Those who did not patronize them

[10]The average alcohol strength of the beer-garden brew is said to be about 3% by volume (Appendix A).

imagined them as dark dens of iniquity and perpetual orgies; but this far outstripped the actuality of customary shebeen behavior. The essential element of the shebeen was that its proprietress, the "shebeen queen," made liquor available to a rather select group of customers in the more intimate setting of a private domicile. Since the largest standard house size available in the townships was only four rather small rooms (two bedrooms, a sitting room, and a kitchen, with an average room size of about 100 square feet), and the house continued to function as living quarters for the shebeen queen and her family (including possibly her husband, her children or other dependents), the space available for paying guests was necessarily limited. Customarily one bedroom was reserved for living quarters, one was reserved as a party room for special customers and could double as a private room in which women entertained their escorts or prostitutes plied their trade, and the sitting room served as the main lounge for drinking and visiting. Although a variety of drinks might be available, bottled European beer was the preferred one.

Shebeens were illegal on the technicality that they sold beer and liquor for profit without licenses to do so. Their operation also violated a more general rule that private dwellings in the townships could not be used for commercial purposes. Occasionally the police rampaged against them, but ordinarily they intervened only when complaints were made. It was generally believed that the police were aware of the presence of most of them, due in no little measure to a prevailing notion among Africans and Europeans alike that the police were aware of everything going on in Rhodesia. In spite of the illegality of shebeens, officials were not rigidly opposed to them so long as they did not provide great apparent disruption. The high price of drinking in the shebeens (up to double the on-sale price at the beer gardens) automatically excluded most of the township population. The shebeens had an unsavory reputation among many Africans who assiduously avoided them and in conversation emphasized their worst aspects: prostitution, sometimes involving young girls; illicit home brewing of strong beverages like "skokiaan," "Tears of the Queen," "kill-me-quick," or "tototo," which were not only unhygienically brewed but were sometimes laced with additives as poisonous as they were potent;[11] and frequently circulated stories

[11] A recipe for skokiaan reported from South Africa was published in an international news release of the Associated Press in 1970: Illegally brewed (native) beer has various names, the most common being skokiaan. Its recipe: two loaves

such as tales of people passing out and later finding that their money had been taken. Others simply did not wish to be thought of as persons of shebeen-type caliber. At the same time, several arguments were customarily offered in support of the shebeens:

1. The shebeens were open at longer and more varied hours than beer gardens and bar-lounges, and thus accommodated people whose work schedules were unusual or, more frequently, who wished to continue social drinking later in the evening than official closing times. One could also obtain "babalazi" or a "regmaker" ("sets you right") in the morning if the only way to overcome that morning-after feeling was to have a drink.

2. Drinking in shebeens was more expensive, and thus patrons tended to be of high status and the atmosphere was usually less rowdy than in beer gardens. An African politician of national visibility interviewed in Salisbury stated that drinking at the shebeens actually saved money for him and for Africans of similar station. In the beer gardens he felt obliged to buy beer for others and was sometimes put upon by the less well-to-do. He was not required to put on such public display of generosity in visits to shebeens.

3. The privacy of the shebeen provided a place where one could take a girlfriend without the danger of word getting back to one's wife.

4. Service was better in shebeens, and patrons were sometimes rewarded with complimentary drinks or coveted snacks like roast meat.

5. Established patrons were able to obtain extensions of credit when short of cash.

6. Juveniles who had the money could drink in shebeens, since there were no restrictions on age. (I would suspect that juveniles were not necessarily welcomed at all shebeens, however, since their presence might threaten many of the advantages which more senior patrons sought.)

Except for the two African men authorized to operate private cocktail lounges in the townships, shebeens provided the only way that individuals could tap the profits of the huge liquor market in the townships. Thus, except in token measure, Africans themselves were

---

bread, one ounce yeast, half pound sugar, corn meal, methylated spirits, carbide, and "other ingredients" to taste. The "other ingredients" sometimes include kerosene or copper sulphate. When acid from a car battery is added the drink is usually called barberton.

prohibited from this source of profit-taking that provided the mainstay of the African Administration. Although shebeens were always managed by African women, they were not exclusively limited to female or even to African interests. One African suggested that "Coloreds, Asians, and Jews" could be found among the promoters who operated in the background and enjoyed a share of the profits of shebeen activity. Several of the bottle stores in town were alleged to cater to the shebeen market.

As was characteristic of most of the "undesirable" facets associated with township drinking, the number of shebeens was typically reported as being "on the decline," although like other problems, this one never quite disappeared. White officials in the Liquor Undertaking estimated that there were "certainly no less than 100 shebeens" operating in 1970. The police estimated about 200 operating on any weekend, noting that a person who decided to operate on one weekend might just as easily decide not to open on the subsequent one. In the impromptu conditions under which many shebeens operated, a precise count of them would have been most difficult and would have been further complicated by the fact that the term itself was generic and not actually used in township protocol. One merely asked, "Where's the nearest party?" An informed African source estimated that there were more than 100 shebeens operating in one township alone, and estimated the Bulawayo total at nearer 450.

The alternative forms of drinking described here have focused on alternatives specifically available to urban Africans. Except for those familiar with the South African scene, the alternative of obtaining a drink in a multiracial setting—particularly the access of Africans to public European drinking establishments—may be conspicuous by its absence. In the ordinary course of events in Rhodesia, one did not see Africans drinking in European bars. In the best hotels and restaurants, Africans were at least theoretically accepted as customers, guided by publicly announced policies that required "standards of behavior and dress which apply to people of all colours."[12] In fact, one seldom saw Africans in such settings. In establishments frequented by the man-on-the-street, such liberal policies and attitudes

[12]Hotels in southern Africa are independently rated on a system of stars from one star (good) to five stars ("compared with the best anywhere in the world"). Only in Bulawayo's best hotel (three star) did I ever see Africans being served in a cocktail lounge, and I observed a refusal of service to Coloureds, persons

were less, rather than more, likely to occur. The little multiracial drinking that did occur in Bulawayo occurred in the African bar-lounges in the townships with limited patronage by a few Whites (Plate 11) or in a small interracial zone between White and Black areas of the city. One old-timer among the pub operators in the European sector of the city volunteered this comment on African drinking:

> "There's no sign outside that says Africans can't come in here. But if they do come in, I get busy with other things, or I just tell them, 'We aren't multiracial in here.' The Prime Minister told us we could have it if we wanted it, but we don't have to have it if we don't want it. In the African townships, they have beautiful cocktail lounges, and you [i.e., Whites] don't go into those places unless they invite you. You see, the Africans don't want it, either. They like to have separate facilities just for themselves."

For the most part, residents of the European sectors of the community had limited information about aspects of township life, much of it garnered from stories told by their cooks and garden boys. Their knowledge of the beer gardens usually centered about opinions related to the great quantities of beer consumed; to newspaper reports providing detailed accounts of incidents, subsequent apprehensions, and eventual trials for stabbings and other altercations occurring in or associated with the beer gardens; and to the great sums spent on beer. Hearing of my interest in the African beer gardens, a young bank clerk on the European side of the city volunteered, "Even with the low wages most of those people make, they sure spend the money on beer. One of our branches over in the townships took in deposits of $15,000 just from three beer gardens this last weekend!" Since beer-garden receipts were handled through Area Controllers and were banked with other monies collected in the townships, one can question the accuracy of such information, but the important fact is the rather generally shared attitude it conveyed: "They sure spend the money on beer."

---

normally recognized as being one step above Africans, in a one-star hotel. Many upper-class Bulawayans insisted that a properly dressed African could go "anywhere in the city." My own observation was that properly dressed Africans did not believe nor invariably find this to be the case.

3

Social Factors in Urban African Drinking

THE TARGET POPULATION of the municipal brewing industry, the African beer drinkers, is the particular topic of the present chapter. But neither this nor any other chapter deals with urban Africans to the exclusion of Rhodesia's European population. The two are interwoven as a unity which reflects, in microcosm, the complex interrelationship between White and Black throughout much of contemporary Africa.

"Social factors" refers to the manner in which people align themselves into social groups and the frequencies and purposes for which those who comprise these groups assemble. The emphasis here is on *who* goes to the beer gardens, as seen in terms of collective behavior. The more elusive question of *why* people go to the beer garden will be considered in a subsequent chapter dealing with individual adaptation. It is important to recognize that who does—or does not—frequent the beer gardens is inexorably bound up with who they are and whom they look to as their reference groups.

Immediately prior to my own fieldwork, rather extensive demographic data had been gathered from a random stratified sample of all the available adults in 192 households (692 respondents in all) in the Bulawayo townships (131). Data collected in the study included basic demographic information and an enumeration of leisure-time activities. Drinkers in the sample were asked where and when they drank and the amount of money they spent. Since the study replicated a survey conducted in Salisbury a year earlier—reported by Reader and May (102) in 1971—comparisons and contrasts between drinking patterns in the 2 communities could also be made. For the summary presented here I have drawn most heavily on data supplied by men interviewed in the Bulawayo survey. Men not only constituted the majority of the beer-garden clientele (averaging perhaps 4 out of every 5 patrons) but were less reluctant to discuss their drinking with field researchers. I have supplemented the data with observations and interviews made during my own fieldwork, including scores of formal interviews with beer drinkers conducted by myself and 2 African assistants.

Bulawayo's African Population

"Don't fall for the commonly held line that the African population of Bulawayo is Matabele just because the city is in the heart of Matabeleland," I was advised by an informed Rhodesian social scientist. "I'd guess that in the townships about 50% are from Mashonaland. As I walk along the streets in Bulawayo, about half of the African conversations I hear are in Shona. And many of the so-called Matabele are really Shona who are acculturated to Matabele ways."

The advice was well given, not because it was any secret that the proportion of Shona (whose traditional homeland is in the eastern part of Rhodesia) plus Africans from tribes outside Rhodesia actually exceeded the number of Ndebele in Matabeleland's major urban center, but because such a fact was at once apparent to any African and relatively unimportant to most Europeans. Tribal variability within the township population provided a major impetus both to form new social alliances and to cling to and even strengthen existing ones. Because urban Africans were housed without regard to tribal origins,[1] meeting places like beer gardens where they could gather voluntarily with fellow tribesmen tended to serve a critical stabilizing function for preserving tribal identities.

Population estimates in 1969 showed an African population in Bulawayo of 190,000, probably a conservative estimate. (Police data

[1]Authorities were not oblivious to tribal origins as much as they were perplexed by what form of constructive accommodation could be made in housing arrangements. An early attempt in Bulawayo to assign residence according to "ethnic grouping" had, as one administrator recalled, been "hotly opposed." Subsequent inquiry in Salisbury has raised the further question of whether township dwellers prefer to live in proximity even to their own relatives (127, *p. ix*). The complex issue of tribalism, including a consideration of the utility of the concept itself, is generally beyond the scope of this work. Reader (101) examines tribalism and detribalization in the Rhodesian setting; Gutkind (45) provides a number of papers discussing the issue in other African contexts. I was intrigued with one African observer's perspective on the alleged detribalizing influence of the city. "We do not become detribalized in the city," he explained. "It's quite the opposite. An African doesn't really have a tribe until he comes here and meets people who are different from him."

in 1971 contrasted an "authorized" population of 170,000 with an estimated one of 200,000.) Of the total urban African population, Africans themselves typically estimated that Ndebele-speaking and Shona-speaking people each comprised "about half" of the total, sometimes adding as an afterthought that the Shona probably represented the "bigger half." The household sample of men aged 17 years and older in the Bulawayo survey produced the following distribution: Shona, 36.8%, Ndebele, 31.7%, Kalanga, 7.0%, and Other, 24.6%. Half of the men interviewed (49.9%) classified themselves as "rural" when asked whether they considered themselves to be rural, urban or semiurban, although the interviews were conducted exclusively in the townships. Only 28% classified themselves as urban; the remaining 22% considered themselves semiurban. Only 7% of the total sample reported being born in town.

About half of the male respondents (48%) reported having attained Grade 6–7 (Standard 4–6) representing the completion of up to 8 years of schooling; 15% had completed some secondary school, 25% had completed Grade 5 or less, and 10% reported having had no formal education. Among respondents of employable status, approximately one in five did not have a job. Those who were employed described a wage range in which most incomes (76%) were between R $4 and $14 per week (modal income was in the category "$6 but less than $8 weekly"). Two persons reported weekly incomes of less than $2, one reported an income over $50, and 22% of the responding men reported receiving no income.

Bulawayo's Beer Drinkers

Demographic data were solicited as part of the University's survey of urban African drinking patterns in Bulawayo. Summarizing the results of the earlier survey made in Salisbury, one observer noted, "Thirty-three percent of the males interviewed in the Salisbury study reported getting drunk every weekend. But, then, there's damn little else to do." Some impressions and comparisions with Salisbury were made possible by the replication study. In general, Bulawayo drinking was characterized by:

(a) A larger number of men who reported that they did not drink (37% nondrinkers in Salisbury, 47% in Bulawayo).

(b) A larger proportion of people who drank African beer (in contrast with the increasing popularity of European beer in Salisbury).

(c) More people who reported drinking daily or during the week rather than only on weekends.

(d) Little use of European-type spirits (only 10 persons in the sample reported using them).

(e) A small proportion (18%) of women who reported that they drank, but a larger proportion among women drinkers (10%, compared with 4.4% of the men drinkers) who drank enough to be classified as "possible alcoholics."

The profile of the "average" drinker in the townships of Bulawayo, based on modal responses, looked like this: He either restricted his drinking to weekends or drank 2 or 3 times a week (bimodal distribution); in either case he did his heaviest drinking afternoons and evenings during weekends. African beer was his preferred drink, and he drank at least 1.5 gallons of it a week (the distribution curve was relatively "flat" through a range of 1 to 8 gallons a week, with a median weekly consumption reported at about 3.5 gallons). He did not get drunk to the point of losing motor control and (naturally enough?) reported that his drinking caused him no trouble.[2] He ate two meals a day and did not substitute drink for meals.[3] His stated reasons for preferring African beer were, in order, food, custom, and price. By contrast, respondents favoring European beer listed, in order, taste, status, and strength, the three complementary alternatives provided in a total list of six. This "average drinker" emphasized what interviewers termed the "physiological" aspects of beer drinking (food, thirst quenching, tastes nice, makes you strong) in stating his most important reason for drinking African beer. He estimated his personal expenditure on beer for himself at below $0.50 per week (median expenditure $0.50–$1.00), and an additional $0.50 spent on buying beer for others, about the same

[2]Only 6.6% of the male drinkers acknowledged "trouble" resulting from their drinking. Researcher Joan May observed: "Respondents were very reluctant to admit that there were any 'troubles' arising from their drinking. Little stigma attaches to drunkenness itself provided that no harm is done, and it is the harm done that is disapproved of, not the drunkenness. But a man who drinks to excess and thereby fails to provide for his family is popularly condemned" (86, *p.* 25).

[3]Eleven per cent of the respondents reported "substituting" African beer for one meal, usually supper. However, an additional 3% reported substituting soft drinks for one meal, and one wonders whether a question about substituting drinking for eating was clearly understood, especially among people who regard beer as food.

amount he received in kind. He did his drinking among friends and at a beer garden, although he reported that he might occasionally drink at his own house or in a cocktail lounge. (Only 2 persons reported drinking in shebeens; one additional respondent described paying for beer at someone else's house. The fact that interviewers were employees of the Housing and Amenities Department may have influenced the range and extent of "outside" drinking actually reported, although the anonymity of responses was assured.) When asked about the main attraction of beer gardens, and provided with a list of alternatives rich with latent sociological hypotheses concerning companionship, opportunities for obtaining home news and maintaining contacts, enjoying the music, dancing and women, or forgetting his troubles, his overwhelming response to the question of why he went to the beer garden was, "to drink beer!"

I will not dwell in more detail on a statistical summary of beer drinking among Rhodesia's urban Africans here, since that reporting has been capably handled elsewhere and represents an effort far beyond the resources of a lone fieldworker. Further, the purpose of the present work is cultural analysis rather than economic survey, and attention is directed more to the ways in which people's lives are variously affected by the beer-drinking custom than to teasing out such statistical estimates as "age-corrected per capita consumption rates." It is, for example, rather academic to report that in 1970 Bulawayo produced enough African beer so that every resident of the townships could have consumed 79 gallons of it. We know that the brewery regularly distributed beer to markets outside the city. We know that many urban Africans neither drank municipally brewed beer nor partook of any spirits at all, whereas an examination of the interview data reveals that some enthusiastic individuals accounted easily for more than 400 gallons a year. Indeed, the interviews suggest that the amount of beer some drinkers claimed to drink may easily have exceeded twice that amount–or even more. But here the concern is with the variability in terms of collectivities of individuals. Who goes to the beer gardens?

Beer-Garden Society

Unfortunately, it is the behavioral scientist's wont to compartmentalize and analyze, and thus in effect to devitalize and sometimes to make more self-conscious the very life processes that attract his attention in the first place. It is necessary here to set about imposing some "order" on the animated aggregate of individuals who constitute the vitality of a busy beer garden and who create so vivid a first impression.

In subsequent sections I shall develop a slightly more elaborate taxonomy for sorting people in terms of their involvement with beer-garden life, but to initiate the process let me employ a simple dichotomy to distinguish between those who come to the beer gardens in the line of work and those who come in the line of pleasure. The former constitute a set of *beneficiaries*, the latter a set of *patrons*. Yet even this dichotomy is misleading in its simplicity, for, as will be described below, one important reason for which large groups of patrons gather in beer gardens is to transact the business of voluntary societies which they have formed in order to secure collective benefits for themselves, even though their apparent presence in the beer gardens is as patrons. I have assigned them a special category as *collective patrons*, or "Clubbers." Just as one man may go to the beer garden because he is sad and another because he is happy, or one goes in order to forget the troubles he has and another goes in deliberate search of new ones, so also do rigid taxonomies designed to provide mutually exclusive categories of drinkers need always to be accepted with the recognition that a man's presence in one role or frame of mind may conflict with another role or frame of mind from past visits. It is even more important to remember that in spite of the universal popularity of the allegation, no one spends all his time in a pub.[4] The social networks that relate individuals and groups within beer gardens are more important in terms of how they relate to elements within the larger society than with how they relate to each other.

[4]I was told that there were cases of older men who did actually "live" in beer gardens, sleeping there after closing time and patiently waiting for someone to offer them beer each new day. However, a beer garden that functions as a "flop house" or a playground or in some other unintended way does so not because it is a beer garden but because it is sufficiently accessible and ambiguous as a physical structure to accommodate other uses and meanings (126, *p. 106*). Since the concern here is with urban drinking rather than with the totality of urban life, this "deviant" and yet, in another sense, "integrative" aspect of the beer gardens will not be explored. Studies in American society have discussed the crucial importance shared by "urban nomads" everywhere of finding a place to sleep (126, 133).

Beneficiaries: The Beer Garden as Place of Business and Commerce

Beer gardens serve directly and indirectly as a place of employment and business for a variety of persons:

1. Managers and Workers. The Bulawayo City Council employed over 500 Africans in the "sales" activity of its Liquor Undertaking and an additional 119 Africans in its brewery and distribution facilities (compared with 14 and 9 Europeans, respectively). Salaries paid by the Liquor Undertaking to African employees were adequate by comparison with factory and clerical wages. Starting full-time salaries were R $25 per month for bar hands. A cashier could earn twice that, and bar managers could earn up to a maximum of R $110. Some officials argued that "everybody wants to get a job in the beer gardens"; others more modestly acknowledged, "It's not great pay—but it's regular. They know we won't shut down, the way factories often do."

2. European Administrative Staff. Although European administrative staff confined their social drinking to other settings, they regularly "spot checked" beer gardens, especially during assigned weekend tours of duty. These checks usually included an "instant inventory" of African beer to ascertain that beer sold plus beer on hand tallied with beer delivered. During such visits some officials made a customary and visible show of camaraderie, including perhaps taking a satisfied sip of beer. It was also a long-standing tradition that administrators assigned to weekend duty arranged to meet at the nicest (and most centrally located) bar-lounge and share a round or two of drinks before concluding their duty day. Their presence in this semisocial gathering, however, was clearly related to their roles as supervisors and inspectors.

3. Police. Maintenance of law and order throughout all of Rhodesia was the responsibility of its national police, the B.S.A.P. (British South Africa Police). The force consisted of both White and African policemen, and their responsibility ranged through town and country, African and European areas alike. Uniformed national police occasionally visited the beer gardens but were not permanently stationed in them. In fact, I was never particularly conscious of a plethora of police anywhere in the country. At the same time, some Africans noted that one could not escape a feeling of "policeness," especially in the beer gardens, where African members of the plain clothes C.I.D. (Criminal Investigation Division) were known to be present and believed to be present in great number.[5]

The municipality also maintained its own small African police force (47 men) whose principal responsibility was to monitor the beer gardens and bottle stores. The presence of these "attendants" (the national police did not acknowledge them as professional policemen: "Let's put it this way—we didn't train them," I was told) lent a note of calm in the beer gardens. The "beer garden police" could detain anyone who became too obstreperous (technically on a "citizen's arrest") until the regular police responded to a telephone summons.

4. Representatives of Special Interest Groups. European and African representatives of the Rhodesian brewing monopoly which produced all of the country's European-type beer visited beer gardens and bar-lounges in a calculated "hard sell" to win converts to their products. Their efforts included distribution of advertising material in the form of printed cardboards that could be folded into sunshades, banners and signs displaying beer labels, inexpensive checkerboards displaying advertising, purchases of beer for clients, supplying barbecued meat at a Braaivelis, and providing entertainment by brewery-sponsored traveling "rock" bands.

European missionaries or African churchmen also made occasional forays into beer gardens to win converts and to distribute tract literature and encouragement. It is interesting to compare their efforts with those of representatives of the commercial breweries, both for similarities and for one important difference. Each group operated on the assumption that a personal introduction to their products by an enthusiastic witness would precipitate a natural desire for more, regardless of any "costs" that might be incurred. However, the brewery tack carefully avoided customary approaches to spreading a message via the printed word. "You just can't reach these people through the usual form of printed advertising," a sales representative explained to me. By contrast, an American missionary said that tracts were eagerly received both inside and outside beer-garden walls: "These people appreciate anything they can get to

[5]The presence of the police, both real and imagined, provoked an ambiguous reaction in which patrons simultaneously complained that there were too many police in terms of interfering with personal and political freedoms, and too few police in terms of bodily protection in the beer gardens and along the routes home.

read." The brewers eschewed both the printed word and the subtlety of soft-sell approaches. Church representatives promised satisfaction; African representatives of the European brewing industry dispensed it.

5. *Individual Entrepreneurs.* Some individuals who came to the beer gardens were in business for themselves. The most talked about in this group were women available as consorts or those hired to induce male customers to continue their evening drinking in shebeens. One young White Rhodesian enthusiastically told me of "a bar somewhere out in the townships—but I'm not sure exactly where, because I've only heard of it myself" where prostitutes "lined up in rows waiting for customers." Like the reputation which shebeens had among Africans, the imagined sexuality surrounding African drinking establishments probably provided soul-food for many European ids.

Myriad hawkers and vendors focused their efforts in or near the beer gardens. Young boys selling fresh vegetables and fruits in plastic bags circulated among patrons. Semipermanently established vendors located adjacent to beer-garden entrances sold a variety of inexpensive new and used items of clothing, shoes and sundries. One could get a shoe shine at many gardens, where shoe shining was operated as an established business rather than the pastime of a horde of youths. Other hawkers moved about selling a host of goods, including artifacts brought in from rural areas, fruit and vegetables, and "hot" items from the European side of town, especially goods acquired by factory workers.

In the Main Beer Hall, an African lessee operated a restaurant with a full range of cooked foods available for 10 to 25¢ per plate. In many of the beer gardens, as well as in the Main Beer Hall, African lessees operated "dry canteens" where candy, cigarettes, soft drinks and snacks could be bought; in some neighborhoods the canteens offered modest mealtime fare and doubled also as a "corner grocery," stocking household staples, fruit and vegetables, eggs, bread, and so forth—with the obvious consequence that young children made visits to the beer garden to purchase treats for themselves or items required at home.

In addition to the food items available at the canteens, a veritable army of independent women food vendors (174 of them according to an informal survey conducted by the Liquor Officer in November

1970) made their daily way to 25 of the larger beer gardens where they sat for hours, legs outstretched before them, beside small cooking fires which marked their occupation and territory. Here they prepared individual food items and inexpensive meals for an assortment of casual and regular food customers, particularly young unmarried men and men whose wives and families lived in the rural areas. Depending on the season, their cooked fare included such items as roasted or boiled meat, chicken, sausages, boiled eggs, sadza,[6] rice, potatoes and sweet potatoes, beans, spinach, cabbage, other leafy vegetables, and the ever-popular roasted green mealies (fresh corn-on-the-cob). Most vendors sold salted peanuts. Roasted caterpillars and mopani worms were also favorites with beer drinkers. Each vendor operated independently of her colleagues. Each had to qualify as a welfare case before being permitted to sell food in the beer gardens; about 40% of the women were widowed or had been deserted. The women reported incomes ranging from $1.40 to $32 a month. Those vendors for whom data were available regarding the number of their dependent children appeared to be a prolific lot, with a median of 6 children each and a total of 538 children for 86 mothers reporting.

On a less visible scale, the beer gardens served as a place for finding clients (particularly for insurance salesmen), for meeting clients to discuss business arrangements, or for joining clients socially at the successful termination of a business deal. An insurance salesman interviewed in Salisbury reported that the beer gardens were an especially good place for him to find teacher clients, an occupational group with a great collective reputation for beer-garden patronage. I was told that Africans in Bulawayo who wished to reciprocate invitations to European associates, but who felt that their own homes were too modest for entertaining, occasionally invited Europeans to be their guests at the nicer of the two privately owned cocktail lounges, where it was possible to purchase both drinks and food. An invitation to go to a beer garden rather than to the cocktail lounge would presumably have been as inappropriate

[6]Sadza is the Shona term for the thick mealie-meal (corn meal) porridge which provides the carbohydrate staple in the Rhodesian African diet. Sadza is usually eaten with boiled meat or boiled vegetables. The Ndebele term for the cooked porridge is *isitshwala*, but the simpler Shona word is heard more often, particularly in conversing with Europeans.

as it was unlikely, since beer-garden patronage was by law restricted to Africans.[7]

6. Birds of Prey. The crowded atmosphere of the beer gardens, particularly on payday Friday evenings, coupled with the slightly tipsy state that many patrons attained, nurtured an interest on the part of some of the townships' down-and-outers who found relatively easy prey for nimble fingers. Although pickpocketing was believed to be prevalent only when large weekend crowds gathered at the out-of-town bus terminal, one of my teen-age informants explained how groups of boys his own age, usually unemployed school leavers, would occasionally befriend an older drinker and drink with him until an opportunity arose for one of them to relieve him of his money. Reports of pay envelopes being lost during Friday evening drinking bouts circulated constantly. Bicycles, purses, wallets and keys were forever being lost or found. Considering the number of unemployed and virtually penniless people in the townships, including young school leavers, it is hardly surprising that many of them migrated toward the beer gardens to see if they could improve their luck. They had little to lose.

7. The Needy. Presumably an even larger group of "have-nots" went to the beer gardens in the hope that someone would buy them beer or share beer with them. No onus was placed on a person who came to the beer hall without money; it was a circumstance everyone understood quite well. Youthful interviewees in the Salisbury study reported their dismay at finding that beer-garden friends deserted them when their own financial luck changed, an ancient bit of knowledge that must be learned anew in each generation. But looking after one's own people is an integral part of the African ethos, and like street beggars who literally earn their doles by faithfully getting to their post regularly each day, the regulars among the poor of each beer garden might be thought of as profiteers who banked their patiently spent hours against the likely odds that sooner or later some beer would come their way.

Collective Patrons: The Clubbers

One of the assumptions that virtually every African I talked with

[7]There was at least one known exception to the African (and Coloured) clientele: a European believed to be of Scottish origin and known as "Tex" was a regular patron at one beer garden. Tex reportedly drank African beer but was not known to purchase it.

seemed to make was that if you were inquiring about his drinking you were interested in aspects of his social life rather than about a physiological and individual confrontation with alcohol. "Alone, beer is never good," one respondent explained. We observed in interviews that the majority of drinkers responded to questions of quantity of beer consumed in terms of the amount consumed by a *group* of drinkers. Our question, "How much beer do *you* usually drink?" was usually answered in the plural, "*We* drink . . . ."

In addition to the ordinary social circumstances of beer-garden drinking, some relatively formal arrangements existed for bringing people together in beer gardens in clubs or societies that are particularly noteworthy. Where to draw the line between a formal and an informal organization is moot. Respondents who reported that they were not members of any clubs or societies nonetheless reported having met with essentially the same group of individuals each weekend for years. Rather than worry over what constitutes a formal club, it is worthwhile to focus attention on the variety of purposes which drew people together for drinking in somewhat formal alliance. Four such types of "clubs" are described.

1. Drinking Clubs. With varying degrees of formality, groups of people who found common bond in their place of origin, in the locality of their urban residence, and in other circumstances in addition to their enjoyment of drinking, joined together in clubs which met regularly at some designated place in a beer garden, most frequently in one of the covered sheds (Plate 8). The clubs purchased and drank their beer collectively from a fund to which each able member made a contribution (of perhaps 20¢ or more per meeting—enough to pay for a gallon). Some clubs required an annual contribution and at some point in the year—usually around Christmas, the most important holiday period of the year for schools and for most wage-earners—unused funds were invested in a party held at a member's house or at the usual meeting place. The tribal drinking clubs, which rotated their meeting place from one beer garden to the next, provided entertainment in the form of tribal dancing and singing for the pleasure of other patrons as well as for their own members, and the drum accompaniment to their gatherings was one of the few sounds that could penetrate the din of animated converation that characterized beer gardens on busy days.

2. Sports Clubs. Sports events, especially soccer and boxing, were a popular pastime among urban Africans for both players and spectators. Weekend events invariably included soccer matches played in one or more townships. Playing fields had been developed adjacent to beer-garden facilities in some areas; in any case, the gardens provided a natural setting in which everyone could gather following a game. The popularity of supporting favorite teams had prompted the formation of organized fan clubs with members paying regular dues. Business meetings of the clubs were not necessarily held in the beer gardens, but occasions for celebration invariably found the merrymakers beating a path from the playing field to the nearest pub. Winning an important game or cup might also be celebrated at a later date by slaughtering an animal in some nearby rural area and enjoying a communal feast and beer drink, the traditional way to validate an important event. The sports societies occasionally expanded their sphere of interest to include the general welfare of their members in the manner of the benevolent societies described next.

3. *Benevolent Societies.* One of the important and striking features of urban African social organization was the way in which people formed new alignments for coping with the uncertainties, particularly economic ones, of urban living. The alleged purpose of such groups was to provide members with a source of aid and mutual support in time of financial stress, but the societies were also social in that they met regularly to conduct their business and occasionally met purely for social purposes (such as a feast planned for a weekend). The societies were not limited to beer-garden associates (churches also sponsored such groups), but the beer gardens provided accessible meeting places and groups organized there did not hold other memberships (e.g., a confession of faith) as a prerequisite. Benevolent activities sometimes extended into related spheres, such as helping members meet rent payments, buy food or meet some unexpected financial crisis, but the reason most extant for organizing was to combat the high costs associated with urban funerals, and thus the groups were known as Burial Societies. These societies were formally constituted and had to be approved by both local and national authority (i.e., the Housing and Amenities Department, and the District Commissioner). One informant who belonged to a church-sponsored society with dues of 25¢ a month explained:

"This is a new thing. It's growing up with urbanization. There's no need for it in the country. Whenever someone died, men went to dig a grave, kill an animal, and drape the corpse. There was no coffin. They just buried people in the ground.

"But here in town, you have to hire buses, pay burial fees in the graveyard, pay for a coffin, pay for a hearse, and buy food for people who come to comfort you or help with the burial. It's quite expensive. Let's say I lose my wife or a child. I might not have the money for a funeral. We earn such little monies that we are not able to save anything. Now if someone in my family dies, I would receive about \$42—or about \$36 if it was a child."

Burial Societies that met in the beer gardens were sometimes described by outsiders as little more than a subterfuge for a weekly drinking bout. My first visit at the regular meeting of one (the Midlands African Burial Society) found, instead, a serious group of 14 Shona men who made a careful distinction between their business purposes and their social activities. Members of the group were all heads of households. They ranged in age from a "youngster" in his mid-20s to one old-timer almost 3 times that age. Their regular Sunday morning meeting was held in a park adjacent to a beer garden and prior to the hour when beer went on sale. The business to be discussed at the meeting was concluded before drinking began, in deference to members who did not drink. Formal dues of \$1 a month (\$3 initiation) were reserved strictly for costs related to the burial of members and their next of kin. Funds available for drinking within the membership and in social meetings with other societies were collected separately, augmented slightly by the assessment of a standing fine of 25¢ levied against anyone arriving late at the regularly scheduled meeting (Plate 8).

Set in a network of reciprocal economic obligations, the beer that most members of Burial Societies enjoyed together socially also provided them with a vehicle for collective fund raising. When a Burial Society found itself in need of extra funds, it could plan a "beer drink" and invite one or more sister societies to the occasion. From its own funds, or from an additional assessment on each member, the host society purchased a large amount of African beer —perhaps 20 or 30 gallons at one time and as much as 200 or more gallons in one day. The gallon mugs were carried to the shelter in which the meeting was to be held, sometimes well in advance of the arrival of the guests. The beer was "sold" to the guests at a profit of 5¢ to 20¢ per mug. The profits remained with the host society—along with an obligation to assume their turn as "guests" when an invitation came from another organization. It is probably these occasions, when the floor of one of the sheds appears to be a mosaic of beer mugs, that has earned the societies the reputation among casual observers of being nothing more than an excuse for a regular drinking spree. Outsiders fail to realize the economic asset that each mug represents or to acknowledge the large number of people seated patiently around the shelter waiting for the somewhat dubious privilege of paying more for a gallon of beer that is warmer and far less inviting looking (as the beer sits, it loses the bubbly head for which it is named and, because of the solids contained in it, begins to settle out) than one which he could purchase more cheaply at a servery only a few yards away.

4. Holisana Groups. A more informal arrangement for mutual help practiced among wage earners was the *holisana.*[8] Like the formally constituted Burial Societies and Sports Clubs, such associations were not dependent on beer drinking or upon beer gardens—one could "holisana" with neighbors, friends or fellow employees—but the convivial surroundings of the beer gardens and the custom that the person who was the beneficiary of the holisana hosted his benefactors on that occasion resulted in the fact that many holisana groups both initiated and regularly transacted their business there.

The purpose of holisana was to help each member, in turn, by supplementing his own limited financial resources with money contributed by each of the other members of the group, whether the group included only 2 persons or as many as 10. At each weekly or monthly payday, one member of the group received not only his own salary but a set sum from the wages of each other member of the group. On the following payday a different member received the nest-egg, and so on until each person received, in a lump sum, the equivalent amount to what he had contributed to every other participant.

Among a money-conscious people whose wage scale was so depressed, the holisana provided not only psychological respite from the doldrums of living "below the bread line" but made possible the periodic acquisition of a sum large enough to make a major investment or to enjoy, at least briefly, the satisfaction of purchasing power. One informant lightly dismissed the whole practice when first describing it: "Personally I don't see any difference, because I still

[8]Holisana=*holisa,* the verb "to pay wages," plus the suffix *-ana* signifying reciprocity.

have to pay back the same amount you have given me." Later he explained that had he not participated in such a group as a younger man he would never have been able to accumulate funds necessary to make the traditional *lobolo* payments to his future father-in-law to compensate for the loss of a daughter and establish a claim to the children resulting from the new union.

In the urban setting, holisana groups sometimes become sufficiently formal and social in nature that the time of payment rather than the payment itself becomes the focus of the exchange. The beneficiary is obligated not only to buy beer for his associates but perhaps even to provide food and drink for them at his home. A new custom was developing in which the social event was the central function of the group, members alternating in serving as hosts to a gathering at which other members were bound to purchase a minimum amount of food and drink. On a more modest but exceedingly widespread scale it was the custom among groups of friends who met informally but regularly at the beer gardens to alternate the obligation to purchase the beer for one day. Thus to the query, "How much will you spend on drinking today?"—a question of interest to officials of the Liquor Undertaking as well as to the researcher—one might receive the reply, "Nothing—it's not my week!"

Individual Patrons: The Beer Garden as Focal Point for Social Life

The drawing power of the beer gardens as a focal point for social life in the townships was attested to not only by the crowds that attended them and the huge quantity of beer consumed but by the fact that beer-garden patrons were not necessarily beer patrons. A few men who reported regular beer-garden attendance and regular spending explained that their money went to buying soft drinks for themselves or buying beer for friends. Men reported that women who regularly frequented the pubs and drank beer seldom if ever bought beer for themselves. Women usually waited for men to buy beer for them. Not infrequently, promiscuously inclined women encouraged some competitiveness among a group of attentive men, the biggest spender winning the company of the woman for the evening—provided, of course, that he was still able to capitalize on his investment.

Beer gardens were largely although by no means exclusively the haunt of men. The men tended to associate with other men of similar origin (language and region from which they came), or type and place of work. Thus men dressed in suits and ties tended to drink together, laborers tended to drink together, older men formed their own cliques and young people—especially if their numbers included boys legally too young to be present—met with persons of their own age in less conspicuous corners and areas. Solitary drinkers were not unknown, and the casual coming and going left others to drink alone during intervals as group alignments changed or people completed their drinking. Still, the over-all tenor of the beer gardens was that of highly social places. It was not essential that someone be talking in each group, but talking and drinking were complementary. A beer-drink was really a talk-fest with intermittent imbibing.

Beer was not drunk quickly. The large plastic mugs were in constant motion from drinker to drinker only if the assembled group was larger than seven or eight persons; otherwise the mug rested in someone's hands for a brief period or was returned to the ground or concrete floor of a shelter after everyone had taken a drink. In a group of only two or three persons, a gallon of beer might be consumed over a period of a half to three-quarters of an hour.

Drinking was usually done while seated, thus emphasizing its leisurely aspects. One rarely saw a person drinking while standing or walking about, unless he had just purchased beer at the servery and had not staked out a location or group of people to join during his drinking. One exception in this regard was pointed out by a bar manager who observed that at the end of the day-shift, when laborers first arrived at the beer garden, a man might buy and "quaff" a half gallon of beer and then purchase another gallon to carry off to join a group of friends in more leisurely fashion. The only individual I encountered who described this practice explained that he did so in order to taste the day's beer before he decided whether or not to stay at that particular beer garden to drink. Beer might be offered to a friend who stopped to exchange greetings and handshakes without his necessarily being invited to join a group on a more permanent basis. Visiting European officials were often invited —or challenged—by the more outgoing patrons to take a drink of their beer. Some officials made a special point of accepting such invitations as a way of giving personal testimony to the purity of the beer.

The vast majority of women in the townships either did not drink or did not admit drinking. In the household survey, 82% of the women interviewed reported that they did not drink (86, *pp. 6, 7*). Women who did drink often noted that they did not join their husbands to do so but drank at other places. Fieldworkers who participated in the survey noted that in households where the husband did not drink wives tended to report that they were not drinkers.[9] At the same time men and women did drink together and some women did accompany their husbands to enjoy either the sociability of the beer gardens or the sociability plus the beer. Salisbury officials had taken a strong stand against allowing pregnant women or women with small children to enter the beer gardens. The Bulawayo administration took the position that children were probably better looked after in the immediate presence of the mother than if left at home, and they did not attempt to exclude infants and small children from beer gardens as long as the mother appeared in control of herself and her children. Similarly they rejected the prohibition against pregnant women on the grounds that Salisbury's only rationalization for the rule was to prevent an expectant mother from dangerous jostling in the crowds; Bulawayo prided itself on having more adequate facilities.

One of the most widely made accusations by critics of the Liquor Undertaking was the youthfulness of many of the persons who frequented the beer gardens and the apparent lack of effort on the part of the municipality to enforce existing regulations restricting underage persons. Some confusion about age arose over the fact that different sets of regulations guided the drinking of African beer and of European liquor. The African Beer Act established a legal min-

[9]Fieldworkers also pondered the possible interpretations which women respondents may have given to interview questions. For example, did a woman who sent to the beer garden for beer but did not herself frequent it consider herself a "drinker?" And what were the complications for an African woman being interviewed by an African man under the circumstances that to be asked whether one drinks was tantamount to an invitation to do so? These and a series of related methodological and substantive issues had led to the formulation of a new dimension of inquiry at the Alcoholism Research Unit into the drinking habits of urban African women, with interviewing by African women. Although that research was just getting under way, preliminary findings suggested that drinking was far more prevalent among urban women than had been estimated in the initial assessment and that, in a pattern of behavior comparable with traditional village life, women beyond child-bearing age were rather free from constraints on drinking. [J. May, personal communication.]

imum age of 18 years, 1 year younger than the Liquor Act which governed the drinking of European beer and spirits for all Rhodesians, White and Black alike. Thus the signs posted at the entrance to every beer garden prohibiting admission to "All juveniles under the age of 19 years" did not apply to 18-year-olds intending to drink African beer. I have already noted that children and youths were present in the beer gardens as hawkers, canteen patrons and onlookers. As one might expect under crowded urban conditions, beer gardens also double as impromptu playgrounds for groups of children in the hours before opening time.

One African youth worker estimated: "Today one is bound to find 30–40% of patrons in the pub under-age. They never do anything to enforce the regulations." Although his statement may have been designed as hyperbole, some modest efforts at head counting and age estimating made in the beer garden where we did most of the interviewing—a stable and settled area with many intact families along with the customary preponderance of unattached males—suggested that if one simply counted all the persons going through one gate over a particular period of time (and ignored their apparent purpose or length of stay), one could substantiate that, at least on Saturday and Sunday, the proportion of youngsters and under-age youth attracted to the beer gardens was substantial. The youth were not "patrons," however; most juveniles in the beer gardens were there as spectators passing through, regardless of how slowly they ambled. I was interested to learn, somewhat belatedly, that in the absence of explicit instructions on my part my assistants "assumed" the 19-year-old minimum in discriminating between adults and juveniles while conducting an age survey. In so doing they confounded my data but illustrated again the ambiguity about just what the minimum age really was for legal beer-garden patronage.

For an American accustomed to the relentless efforts of state authorities to enforce liquor regulations (particularly in regard to under-age drinking) and to the practice of placing the burden of proof of legal drinking age on the youthful or youthful-appearing patron by insisting that he be able to produce identification ("I.D." card) as proof of age, the solution to the under-age drinking problem seems obvious but suffers from being somewhat unrealistic. Few Africans carried identification; those who did were customarily older people in responsible jobs. Only a fraction of urban Africans possessed a much-coveted driver's license. Still fewer possessed passports or other identifying documents. Municipal police occasionally assigned to monitor beer-garden entrances had to make their estimates on the basis of looks and the intensity of the verbal affidavits of a youth's associates. During the crowded times (the very times that made the beer gardens most entrancing), while one teen-ager might be arguing his age with a security guard ten others could slip by and lose themselves in the crowd. But teen-agers did not need to enter the beer gardens in order to obtain beer—their older friends went in to buy it for them legally and carried it outside or handed it over the walls. On a pleasant weekend some beer gardens had a number of satellite "gardens" in operation nearby where groups of boys and girls, of ages reportedly as young as 12 years, gathered to emulate their seniors. Official concern was coupled with a considerable uncertainty regarding what to do about the problem without creating unnecessary new ones. Officials took some solace in the fact that the beer was low in alcohol content and that most young people had limited funds to spend.

Alternative Pursuits to Beer Drinking

Demographic data already reviewed suggest that a significant proportion of the township population did not frequent the beer gardens. What did these people do with their leisure time?

In earlier sections I have discussed some alternative recourses for obtaining liquor, such as drinking at private parties or frequenting shebeens and bar-lounges. One alternative for obtaining beer—returning to one's home in the tribal trust lands—was an important leisure-time alternative for nondrinkers as well as for drinkers. The weekend buses (and, to a lesser extent, 3d- and 4th-class coaches on the trains) transported hordes of people who left the city as soon as possible at the end of the work week and returned late Sunday or early Monday morning. To a less dramatic extent than the weekend exodus this urban–rural migration was evident at all times among women, unemployed men and employees on holiday. At Christmas many factories closed for 2 or more weeks; European employers took family holidays to coincide with the annual school calendar, and a major exodus occurred among Africans headed for "home."

For those who did not, or could not, escape the city so readily, the alternatives to the social life of the beer gardens were limited. The formal recreational facilities available were at once a tribute to the determination of the Housing and Amenities Department to make townships liveable and mute testimony to the unlikelihood of ever providing adequate facilities for an exploding urban African population. The combined township population shared 11 tennis courts, a 9-hole golf course, and 2 swimming pools. Film shows and ballroom dancing were sponsored at 8 community centers. Weekend sports events drew large crowds of spectators. The popularity of soccer was evidenced by the provision of 2 main stadia with grassed fields and 2 dozen other playing areas located throughout the townships, although it was pointed out that the maximum crowd at a soccer game ("About 6,000 if we close up the other games") was comparable to the number of persons attending an average beer garden on a weekend. Additional sports facilities included a gymnasium a regulation athletics track, a bicycle racing track, 2 basketball courts, an outdoor boxing stadium and a tribal dance arena. There were "club" facilities, youth centers and libraries. The Housing and Amenities Department had also encouraged the development of Guilds—voluntary "adult cultural associations" organized at a number of community centers to sponsor programs of entertainment and recreational activity. Interest was reportedly high in social dancing, various music activities and tennis, but total Guild participation had remained small. Church attendance at a great variety and number of churches throughout the townships was large. Churches also sponsored activities which engaged volunteer members for other hours on Sunday and during weekday evenings.

Questioned about their leisure pursuits, 31% of the men interviewed in the Bulawayo survey stated that they most liked to spend their time in the beer halls (86, *p. 34*). The remainder indicated preferences for alternative forms of leisure in the following order: watching sports (18%); home pursuits (gardening, etc.) and meeting in church groups (15% each); listening to the radio (6%); intervisiting and attending other organized groups (5% each); playing in sports (4%); going to night clubs (2%).

Attending the cinema, either at theaters in town or at inexpensive showings in neighborhood centers, was not stated as a first preference by any of the 411 men interviewed, a pattern which several Africans explained as a consequence of the "third rate" films usually shown and the corollary that movies had come to be thought of as a form of entertainment for children and for the unsophisticated. Television viewing on Rhodesia's one channel occurred rarely among township inhabitants but was generally looked upon as a very desirable pastime by anyone who had access to a set. Hire-purchase schemes put television sets within the range of families of moderate income, but the aerial forest of television antennas characteristic of the urban American ghetto was absent in the townships.

Reading was not included among the alternatives posed on the question about leisure preferences in the township survey. My own questioning and interviewing suggest that reading was a frequent pastime, especially among older teen-agers who almost invariably had completed at least 8 years of schooling. A great deal of the reading reported by this group was done in the pursuit of correspondence courses and the preparation for examinations for nonresident students at upper secondary and advanced levels. I suspect that the compulsiveness that surrounded the tendency to associate reading with "improving oneself," coupled with the solitary nature of the reading act, precluded it from being identified as a "leisure" activity. Readers sometimes reported looking at "comics" and reading novels. Most, however, seemed totally instrumental in nature. Young "school leavers" (in the sense of having been forced out in the fierce competition for places in African schools beyond primary level) in unknown but presumably great numbers pored over their own and borrowed books for several hours of self-imposed study each day in the hope that the next level of "education" would provide the necessary magic to get them a job. Librarians in the townships reported difficulty in keeping books on their shelves because of the great demand. One library report noted "a marked preference for nonfiction books, history, geography, literature and science being the most popular subjects." The fact that reading was regarded as something one had to do meant, for many of the young unemployed, that a period during which one engaged in the solitary act of reading provided a good reason for then going off to join friends at the beer garden. One 18-year-old boy explained to me that although he did read occasionally, reading was never his first choice. "Going to drink with my friends is more fun than reading."

4

Cultural Factors in Urban Beer Drinking

ALTHOUGH ANTHROPOLOGISTS have long argued over the most suitable definition for explaining precisely what culture is (65, *Ch. II*), they generally define it to include the totality of learned, socially transmitted behavior—everything that is suggested by the frequently heard term "custom." Further distinctions, and important ones here, are the different levels of abstraction by which we refer to "culture," viewed collectively, such as the traditional culture of the Ndebele or Shona, and a "subculture" or a cultural system, such as the shared and socially transmitted set of behaviors one might observe among a group of people who meet regularly to drink together in a beer garden. The present chapter examines urban beer drinking in Rhodesia from the cultural perspective. Comparing and contrasting statements are made about traditional and contemporary drinking and about African and European drinking. Beer-garden life is then examined as a subculture of its own.

Beer in Traditional Culture

In reviewing accounts of traditional Ndebele and Shona life (20, 24, 50, 71, 89, 92, 135) as well as accounts of other southern African groups (e.g., 1, 26, 104) one is struck by the frequency with which the brewing and drinking of beer is mentioned in relation to social interaction. One sometimes gathers the impression from such accounts, and from the comments of first- and second-hand informants, that the men "just sat around drinking beer all day." Such a portrayal seems as unlikely an accurate description of beer drinking in the past as it is of the present. At the same time, it suggests the pervasive nature of beer drinking in traditional African societies.

I have not attempted to present an inclusive review of traditional drinking. I do wish to call attention to the two major social purposes for which beer was brewed and to discuss each of them (with illustrative examples from the Ndebele) sufficiently to provide a perspective for recognizing the link between traditional and contemporary drinking. Were traditional drinking the focus of the present study, a major effort would have been necessary to uncover and catalog all the purposes for brewing beer and the circumstances for drinking it. I often found my bibliographic resources and informants presenting different and conflicting facts and opinions about traditional drinking. I have assumed that each source has something to contribute to the complete account; only a few important features are presented here.

One key informant suggested that traditional brewing could be described in terms of a natural division between beer brewed for *work* and beer brewed for *ceremonial purposes.* This distinction is followed here. That more attention is devoted to describing ceremonial purposes does not mean to suggest that such purposes actually predominated in daily life (much the same as those of us who enjoy drinking might be able to suggest a number of occasions when drink is especially appropriate but are also willing, as Cervantes once noted, to drink "on no occasion at all"). The variety of ways in which ceremonial drinking was interwoven into the fabric of individual and collective life is actually of less consequence than the fact that drinking was in some way linked with such occasions. As I have already noted, ceremonial drinking is a crucial aspect of "Integrated Drinking"; its presence in traditional culture provides another integrative dimension in the continuity between past and present.

I was curious to learn how my informant could link all ceremonial drinking to pleasing or demonstrating respect for the ancestors (*amadlozi*) when some of the occasions he described seemed so blatantly and expressly designed for the satisfaction of persons immediately at hand. "I'm not always sure myself just how these things are linked up with the ancestors," he explained. "The old people have been reluctant to tell us about the old ways. They feel we might let it be known to the Whites. But in general you may say that sometimes the spirits are happy because they see that the people themselves are happy, even if the beer has not been brewed especially in their honor."[1]

[1]From fieldwork among the Shona, Murphree (89, *p. 58n.*) has provided an explanation offered by an informant of how beer can simultaneously be offered to the spirits and consumed by mortals: "Once, when during a *rukoto* ceremony [conducted for rain-making] the sacralized beer had been presented to a number of *mhondoro* [tribal spirits], each named in the prayer, it was immediately

The occasions for calling an *ilima* or work party were numerous, for traditional life tended to be communal, and any task that required or could be facilitated by a collective effort was customarily approached by inviting one's neighbors for a beer drink, an implicit request for assistance. Plowing, building or repairing kraal fences, threshing corn and constructing a new hut were all occasions at which a work party might be called. It should be borne in mind that from the host's point of view the work was the focus of the gathering, the beer a stimulus to drawing the group together. Similarly beer was said to be provided occasionally during an *inkundla* (hearing of a case by the elders and headman) or an *indaba* (a high-level discussion of important affairs), although apparently some elders preferred to delay the drinking at such meetings until the business at hand had been settled soberly, a concern still expressed among and about African politicians in the contemporary scene.

Ceremonial beer drinking occurred more frequently as a result of events in individual life cycles than in events shared simultaneously by people living in a wide area, presumably as a consequence of both the small-scale groupings in which people lived (59) and the fact that beer brewing was itself a small-scale process. Communal celebrations of large scale did occur. Celebrating the victorious return of the Ndebele warriors, for example, was a not-infrequent occasion among a people whose warring activities had such an influence on the modern political boundaries of the subcontinent (19, *p. 1*). An annual celebration of dancing and feasting (*inxwala*) was another ceremony observed among the Ndebele as a nation rather than in their separate tribes.

One important collective use of beer was in propitiating the Supreme Being in whom the Ndebele believed. Although the names by which their deity has been known, the manner in which he has

---

ladled out to the participants and observers present. I asked the medium whether the *mhondoro* would not be angry that we had drunk the beer designated for them. He smiled and replied, 'We do not believe that this beer is actually drunk by the *mhondoro*. But the *mhondoro* are here, and they do drink it in a spiritual way. If, for instance, in the evening your wife places many appetizing foods on the table in preparation for breakfast and you dream that night of eating them, does that mean that when you wake in the morning they will not be there? But did you not really eat them in your dream as well? Just so, the *mhondoro* drink as we would in a dream, and they are glad that we drink with them here in a different way.' "

been honored, and the extent to which the ancestors were called upon to act as intermediaries have constantly undergone change (24, *pp. 72–76*), specific calamities and generalized anxieties continue to be met by asking the deity for help. In this regard, rain has always been the lifeblood of the Ndebele; traditional rain-making observances are still practiced (27). Beer was specially brewed for these occasions and surrendered to one individual responsible for both conducting a formal ceremony in order to assure a needed rainfall and for subsequently learning in the course of the ceremony whether special practices, such as plowing only on certain days, would have to be observed. For this occasion beer was not only consumed by the participants but was also deposited in pots outside one of the caves where the Supreme Being was said to reside.

Individual or personal occasions for brewing beer and calling relatives and friends together to drink it were particularly associated with life-cycle events such as the following:

1. Birth of a first child, particularly if the parents had been childless for a long time.

2. The "beginning of womanhood" of a daughter, an event which was socially validated by a series of feasts attended by relatives and friends at the end of a period of seclusion for the girl herself (24, *p. 53*).

3. Weddings, for which beer was brewed at each of several steps in the negotiation and ceremonial proceedings.

4. At the time of paying *lobolo,* a payment made by the suitor to the bride's father. Traditionally such payments were made when the birth of a child consummated the union; more recently, the payment has usually been demanded before the marriage ceremony.

5. At a burial and, a few weeks later, at a ritual washing in beer of the implements used to dig the graves.[2]

6. Most important, at the *umbuyiso* ceremony, held (ideally) 1 year after the death of a male head-of-household in which, through a ceremony complementary to the separation rite of the burial, the spirit of the deceased was incorporated back into village life, taking up residence in a living ox, *idondolo.* Great quantities of beer were provided at this time. Token amounts were sprinkled on the ground to induce the spirit to enter the village, and offerings of beer and meat

[2]This is the only situation I have found in which beer served a symbolic function in purification.

were made to him. The spirit was told that he had been brought back home to look after the family. Although the *umbuyiso* ceremony was performed only once for a deceased man, at subsequent occasions respect was shown to particular spirits by holding sacrifices and feasts to honor them, and at such time the new head of family poured beer over the animal in which the spirit was now residing, for the welfare of the living depended on the well-being of the spirits, just as the spirits in turn depended on the living to honor and respect them.[3]

There were still other occasions at which beer was brewed and consumed. Beer was sometimes brewed to honor a guest. It was also brewed if an *inyanga* (African curer) had been called to a village to cure an individual or to protect a home to make it strong enough to ward off the constant danger of evil spirits or wizards. And, in addition to the host of pragmatic or ceremonial reasons described above, beer was also brewed when a man felt like having a good time drinking with his kinsmen and neighbors. The occasion was noted by a term descriptive of the social context in which it was to be consumed, *mhlalaphansi*, "the beer we drink sitting down."

One finds, in reviewing the ethnographic literature on drinking, that in any society in which the use of intoxicating beverages is incorporated into secular life, there have been those whose drinking obviously exceeded accepted group norms. Based on the cross-cultural study described earlier (25), Barry (4, *p. 460*) has noted that reports of addictive drinking were extremely rare in the ethnographic accounts reviewed and that most of the alcoholics who were reported were chieftains who had access to more liquor than did commoners.

Popular accounts of the brief royal lineage in Ndebele history lend support to this observation by suggesting that excessive drinking may have been a problem in the life of one or both of its two kings. One author, Mhlagazanhlansi (Reverend Neville Jones), portrays a decline in the royal lineage by contrasting the attitude toward drink exhibited by father and son. Mzilikazi, the first king of Ndebele, "was definitely set against everything calculated to lower the physique or undermine the morale of his people and, in this connection, he refused to countenance excessive beer drinking and promiscuous sexual intercourse. Only the adult men were allowed to drink beer,

[3]See Child (24, *pp. 61–64*). For comparable ceremonies among the Shona, see Holleman (51).

and it was not until the youths had become men by assuming the *inkehla*, or head ring, that they were allowed to marry" (88, *p. 17*). On the other hand, Mzilikazi's son Lobengula, the second and last in the royal succession, "repealed many of the wise laws made by his father for the purpose of maintaining the morale of his people, and permitted the soldiers to commit all sorts of excesses. Widespread beer drinking and licentiousness, such as would have been suppressed by Mzilikazi with a heavy hand, were allowed. He opened the door to weakness in the nation" (*p. 18*).

Elsewhere a different picture is given of Mzilikazi, at least near the end of his days: "All contemporary accounts during the last fifteen years of his life lay stress on the enormous quantities of beer the king consumed, and it was inevitable that he soon began to show signs of chronic alcoholic poisoning. . . . His interests narrow down to the presents that can be extracted from his European visitors and the pots of beer which are passed to him almost continuously by a circle of obsequious queens" (100, *p. 116*).

In reviewing the pervasive role which beer played in traditional native life, it is important to note the division of labor between the sexes and related differences regarding brewing and drinking. Men and women did not pursue the rounds of their daily activities in the company of the opposite sex. The men's activities were performed with the assistance and counsel of other men; boys were responsible for the endless task of herding livestock; and women's activities were performed among older women, older girls and young children. The men did some of the heavier work, but the women's domain extended beyond the kitchen and care of small children to caring for the fields as well. It was not customary for the men to look after the children or to spend time around the home. Men simply were not part of the hour-by-hour domestic life of the household. If this tradition continues to exert an influence on the behavior of urban men today, and male informants suggest that it does, then the beer gardens fulfill an important role by providing a place for men to gather away from the feminine sphere of hearth and home. In the infinitely more crowded atmosphere of the townships and in the tiny dwellings and plots which have replaced his traditional kraal, the man who elects to stay at home or who buys beer and returns home to drink it has adopted a markedly different alternative to traditional male behavior in spite of circumstances that would seem to weigh heavily in favor of observing former customs.

Contrasts between Traditional and Urban African Drinking

Traditional brewing was the responsibility of the women. A man directed his wife to brew beer; he did not brew it himself. And though one always knew who had brewed the beer one was offered, the host drank first in affirmation of the quality and purity of the beer. Initially the local municipalities followed the custom of employing African women to brew their beer,[4] but problems of both quality and quantity led inevitably to the modernization of the industry. To achieve the level of production and technology necessary to satisfy present demands, the municipal brewery violated virtually every canon of traditional brewing except that beer was still brewed in batches rather than by continual process. Beer was brewed and kept in metal tanks rather than in clay pots. A variety of new grains such as barley was utilized in brewing. The brewers were not only men (rather than women) but were strangers as well as Whites. Beer was brewed with what Africans referred to as "chemicals and other things," a practice almost guaranteed to arouse suspicion. Finally, except for quality control, the brewers did not drink the product they brewed (the chief brewer was, in fact, a teetotaler). What was more, they brewed beer explicitly for profit, a custom which had never been prevalent and which was regarded as unlucky in some rural areas.

The drinkers themselves had also changed. Traditional drinking was essentially a pastime of men, reserved especially for the old men but extended to "younger" men (of approximately mid- and late-30s) established as heads of households. Younger men did not customarily join in a pastime reserved for their seniors unless specifically "called" by an elder to have a drink. Women reportedly drank together in groups apart from men and they generally drank less beer or consumed nonalcoholic drinks.[5]

[4]Officials in Bulawayo sought to honor their earliest municipal brewers by renaming an old beer garden with the name of one of the best-known among them, Madlodlo. The unveiling ceremony marking the opening of new facilities at the garden in November 1970 drew an amused reaction among the Africans present that seemed likely to continue to be associated with the new name. Perhaps few of the Africans present knew who Madlodlo was, but they recognized that it was not an African practice to name places after women, and certainly one did not name beer gardens after them.

[5]This impressionistic stereotype of drinking by women is difficult to docu-

Urban beer-garden populations included their coteries of older men, but the beer-garden clientele seemed to tolerate rather than to exalt its senior patrons. The old men frequented the gardens, drank the commercially produced African beer, and occasionally drew groups of younger men interested in hearing their stories of the past. Beer gardens tended to be robust and vital, dominated by younger men seeking the companionship of their peers and, to a lesser extent, the company of women. Among the younger crowd, beer was a focus of attention rather than merely an adjunct to conviviality. Their drinking was more self-conscious, and one was aware of status differences being displayed through purchases of different brands or types of beer and of the presence of hangers-on whose motivation was not companionship as much as a calculated ploy to cadge a beer from someone more affluent.

The motivations of the urban drinker had also changed. If traditionally he drank in celebration, now he found less occasion to celebrate. His work was never "done" in the shop or factory unless he had been dismissed or laid off, in which case he had little cause to celebrate as he joined the masses of the unemployed. The receipt of his weekly or monthly pay envelope marked only a momentary respite in his unending round of immediate expenses, and the amount of his pay, contrasted with the costs of rent, maintaining a family, paying off loans, meeting school fees, and so forth, served to emphasize the near futility of his endeavor. His industrial circumstances offered no possibility for the windfall of an unexpected "good" year, the one sustaining hope in an agrarian lifestyle.

The drinking patterns of the urban dweller, like most of his other social activities, were regulated by the hour of the day and the day of the week. The African urbanite found himself drinking a little each day or at a regular time each week as he learned to suit the hours of his desire for beer to the hours when he could purchase it. His counterpart in the tribal trust lands, although he could not obtain a drink unless he or others had beer on hand, was at least free to arrange for brewing to begin anytime and to anticipate the brew he had commissioned, and once his beer was made and the drinking

ment. During my initial opportunity to observe beer drinking in the reserves, the first gourd of beer I noticed was handed to an old woman by an old man seated next to her in the shade of a hut. Nevertheless, the written evidence and recollections of informants do suggest that traditional drinking was an activity pursued essentially by older men but not restricted to them.

begun, there was no enforced "closing time" until the beer or the drinkers themselves were exhausted (Plates 2 and 3).

Municipal beer itself provided further contrasts with its traditional counterpart. The enthusiasm and concern which many city fathers and administrators exhibited about brewing and distribution aroused suspicion. Was the beer a thoughtful and humanitarian attempt on the part of the European community to make urban life palatable by providing a reasonable and refreshing respite from it, or was the beer part of a conscious plot to keep the Black man insensate so that his status could be extended into perpetuity? The latter possibility nagged many African drinkers. The charge would have been incredibly difficult to prove, but it was equally difficult to disprove. Everyone knew of individuals from whom drinking was taking or had already taken an obvious physical toll. Such generalized and pervasive suspicion was not thinkable in the days when brewing was an individual matter. Fear of individual retribution, effected either by psychological means or by direct ones, has always characterized African belief systems and patterns of interpersonal control. No one, however, would have been suspected of questionable motives involving the entire African nation.

Municipally brewed beer aroused still other kinds of suspicion. Employee attempts to dilute it with water had always presented a problem. The consequence was that everyone connected with the municipal beer industries was suspicious of everyone else. Officials in the Housing and Amenities Department found themselves having to protect beer brewed for Africans from attempts to adulterate it on the part of other Africans. Efforts at vigilance were complicated by the fact that at times batches of beer varied in taste, depending particularly on variations in available grains and on the length of the fermenting period. Beer held longer tasted stronger, since it continued to ferment. Thus the last of one day's delivery might differ markedly in taste from a fresh delivery the next day. Home-brewed beer had always been subject to differences in quality from one batch to the next, and women who brewed consistently better beer than their colleagues were recognized and respected for their skill. But on the scale that the municipalities marketed their beer, patrons expected and demanded consistency. Perceived differences in beer from one delivery to the next aroused protests and accusations that the beer was not up to standard or had been adulterated. It was to this point that a beer-garden patron was speaking when he proclaimed in Ndebele to his colleagues a comment clearly intended for my ears as well, "When the beer is good, we rejoice. But when it is bad, we complain bitterly" (Plate 5).

Real and imagined variations in beer quality had sometimes resulted in unpleasant rumors, temporary customer boycotts and even mob reaction. Laboratory tests conducted at the brewery had disclosed the somewhat disconcerting fact that up to 10% water could be added to the beer without the fact being established beyond doubt (18, *p. 1*). Officials reassured themselves that if African beer was so amenable to watering the practice of diluting it must have gone on as long as brewing itself. African informants were not swayed by this argument. They pointed out that a host brewed a substantial amount of beer for any occasion at hand but would never have diluted it to make it go further. Running short of beer attested to one's popularity and standing in his community by demonstrating that in spite of brewing a great quantity of beer he had so many friends that he could not possibly brew enough for them all.

The municipalities had not been reticent in boasting of the quality of their brew and its likeness to traditional beer. They carefully tagged their products with descriptive names from the indigenous languages. Bulawayo's beer had Ndebele labels. The draught beer, *Ingwebu,* called attention to the frothy head on the beer to suggest its liveliness. One of the two packaged beers was called *Jabulisa,* happiness; the other, *Indlovu,* the Ndebele word for elephant, at once suggested strength and was linked to the symbol of an elephant appearing on the city's official crest.

Salisbury provided Shona terms for its four products: *Rufaro Beer, Rufaro Ngoto, Rufaro Mhamba,* and an obvious newcomer, *Rufaro Sportsman.* That municipality offered the following explanation of its brand names (111, *p. 17*). " 'Rufaro' is a Shona word, which means 'Happiness.' 'Ngoto' is best translated as 'The strong beer as brewed by the Elders in the Village (Kraal),' and 'Mhamba' simply means 'Beer.' "

An element of particular interest to the anthropologist in comparing traditional and urban drinking practices is the extent to which municipally brewed beer had been accepted as a substitute for traditional beer in ceremonial as well as in secular use. Such acceptance provided a pragmatic test of whether the urbanites consumed municipal beer because it was inexpensive and the only legally available drink they could buy or whether they had come to accept it as the contemporary equivalent of their traditional brews. My too-obvious interest in this point stirred up more commotion and controversy than I had intended, and I found both African and European informants taking and rigorously defending conflicting points of view. European officials preferred to believe that their commercial product enjoyed such acceptance that it was quite acceptable for ceremonial purposes within the townships, although they usually recommended checking with an African to verify their impression. Africans whose duties brought them frequently to Head Office were sometimes equivocal in their explanations, caught between the custom and courtesy of giving European officials the answers they believed were expected and the possible accusation of showing a total disregard for the truth. I was told how the municipal brew might be acceptable to the ancestors if one "talked it over with the spirits beforehand" and had been able to explain to them the difficulty of obtaining properly brewed beer. On pressing the issue, I learned that a more customary resolution of the dilemma was to brew a small amount of beer at home—perhaps only 4 gallons, not enough to call attention or even to get one in great trouble if he happened to get caught—and to offer this beer to the ancestors while the mortals among the honored guests were presented with draught beer purchased at the nearest beer garden.

Other Africans scoffed at the idea and insisted that it was utterly impossible for municipal beer to be used for ceremonial purposes. Ceremonial beer was not only *offered* to the ancestors, it had to be *brewed* especially for them, and the ancestors were informed at each step of the process. One informant explained:

> "I think it is very unlikely that municipal beer could be used. If a family wanted to brew beer to appease the spirits—perhaps because of illness in the family—they have to speak certain words before the corn is soft. 'Here is the corn to brew *your* beer,' they would tell the spirits, even before they begin soaking the corn."

Another informant drew an analogy between using commercially brewed beer for ceremonial purposes and going to the butcher shop to buy meat for a ceremonial feast:

> "You have to talk to the beast *before* it is slaughtered if it is to be used in a ceremonial feast. And the malt has to be spoken to before it can come ready to brew, the same as an animal. That's when the ceremony really

starts. So from what I gather, it would not be possible just to buy beer and use it."[6]

Further investigation led me to realize that everyone was portraying an accurate point of view from one perspective. The divergence was due to the dynamics of the situation itself and was linked to variation in degrees of urbanizing among individuals in the townships and to their opportunities for contact with persons residing in the country. The preferred mode was to conduct ceremonial activities in the tribal trust lands where activities could be carried out in the traditional manner. But for people who had lost contact with their rural communities, or for those consciously trying to shed that heritage (yet not so bold as to risk offending the ancestors), municipal beer was used or was deemed to be a possible substitute. One-third of the persons interviewed in the beer gardens said that they had heard of cases where municipally brewed beer had been used in the important umbuyiso ceremony. Although one must allow for the likelihood of acquiescent responses, my assistants and I were satisfied that we talked with at least some persons who had firsthand experience at such occasions. Explanations volunteered by those interviewed ranged from the urbanite's lack of alternatives—"My life is here in town. This is the only beer we can get"—to complete empathy with tradition: "You can't use this kind of beer. It would be against the ancestral spirits. You've got to *brew* the beer, not *buy* it."

Contrasts between Urban African and European Drinking

"Shake well" advises the printed decoration on the outside of the waxed carton of packaged African beer. The European observer recognizes at once a small but dramatic difference between the caution he exercises not to excite the easily aroused contents of his bottle of beer and the necessity of agitating the solids that tend to settle in the bottom of a package, mug or tank of African beer. One responsibility of each beer-garden manager was to see that the contents of the huge 500-gallon tanks were regularly stirred whenever beer was actually being drawn from them for sale.

The practice of passing a plastic mug from one drinker to another, rather than for each person to drink from his own container, also

[6]A reader of an earlier draft pointed out that although words are spoken over the grain or animal, they are actually addressed to the ancestors being honored and not to the products intended for consumption.

strikes the fastidious European. His own beliefs about "germs" may have a subtle but important impact on his initial reaction to African beer, although he may display his reticence by directing his comments toward the taste of the beer rather than to the dubious hygienic circumstances under which it is consumed.

The European drinker—or the barman or waiter serving him—carefully satisfies himself that the last drop of beer has been poured before removing a bottle from bar or table, since the last drop is as clear as the first, and the beer belongs to the customer. The consumer of African beer is not so insistent about "finishing" the last drop of his beer. The suspended solids begin to settle as soon as the beer is served. Most drinkers leave the thick portion which settles at the bottom of one container and turn instead to the frothier top of another one. Those few who actually do empty their mug, a practice usually followed only by the very old men, explain that the thick part of the beer provides them with "food."

The African makes no attempt to repress the first indications of tipsiness—if he feels relaxed, he wants to look relaxed—in contrast with a middle-class European counterpart who may insist as he leaves the cocktail party that he is "as sober as a judge." Judicious as the middle-class White's drinking behavior may seem in comparing stereotypes, one can nonetheless identify certain characteristics of European drinking that are rare or absent among African drinkers. Among the latter there appear to be relatively few alcoholics. Less than 4% of the male drinkers, or 2% of the total adult male population, was classified as "alcoholics" in the Bulawayo survey (86, *p.* 23). Some other antisocial aspects characteristic of White middle-class drinking were also rare or absent: hurried drinking, solitary or hidden drinking, and personal anxiety about drinking. African beer drinking also differed by extending into sacred as well as social domains.

The American observer is struck with another contrast, the emphasis placed on the nutritional value of African beer. American advertising stresses the context in which beer is consumed—the enjoyment, the sociability, the refreshing qualities—virtually everything about beer except its high caloric value.[7] Americans are barraged with

[7]American advertisers are, in fact, prohibited by law from making claims concerning any health-sustaining properties of their beers, such as caloric or nutritional value or vitamin content. However, one manufacturer that produces a beer (Gablinger) containing fewer calories is apparently allowed to advertise that fact.

advertising slogans that concertedly ignore that dimension: "The beer that made Milwaukee famous"; "Beer belongs—enjoy it"; "The beer beer-drinkers drink"; "Blue ribbon"; "High life"; "It's the water"; "Mabel, Black Label"; and "Bud—that's beer." Calorie-conscious Americans would be shocked by an advertising campaign heralding the one attribute most often ascribed to African beer—its value as a food:

> "It is mainly to these solids and high protein content that African beer owes its great nutritional value. One gallon of good African Beer contains approximately 2,800 calories and about 26.6 grams of proteins, as well as having a high vitamin content" (110, *p.* 5).

I recognize that my reaction may have been peculiarly American and that emphasizing the food value of an alcoholic beverage did not seem untoward to Rhodesians of either African or European origin. One international authority on alcohol has suggested that Europeans are as surprised at the extent of product advertising common in the United States to encourage the sale of alcoholic beverages as Americans are surprised to observe how European countries "extol the nutritional and medicinal properties of such beverages" (63). "Guinness is good for you," proclaimed a sign in a European pub in Bulawayo, and although most Rhodesians had not seen a pint of Guinness since the beginning of the sanctions that followed on the heels of "U.D.I.," the slogan revealed a characteristic European attitude toward the benefits of drink.

References were made constantly to the nutritional value of African beer both by those who sold it and by those who bought it. The drinkers themselves did not categorically attribute health-sustaining properties to all alcoholic beverages, however. Several of those with whom I talked casually or held a formal interview, especially the younger men, expressed concern with the drinking of "strong stuff" (brandy, gin or cane spirits in particular), stressing that it was not only unhealthy but hazardous. Only nine men (and no women) interviewed in the household survey in Bulawayo reported drinking spirits at all (86, *p.* 6). However, some Africans had developed a taste for such drinks and drank them when finances permitted or bought them in small amounts (the most frequent purchases of spirits in the municipalities were in the smallest units available, the "quarter" or "nip," one-fourth of a quart, and the still smaller "miniatures" available for 30¢ at the beer gardens) and drank them straight or used them to lace their beer to make "a strong drink for a strong man." Draught beer was also laced with other products, the preferences for such additives apparently reflecting some age-graded patterns. Men and women sometimes laced their beer with a quart of "stronger" packaged beer; juveniles reportedly added sugar, Coca Cola, powdered milk, and even ice cream to theirs.

My own preference in drinking is for distilled spirits mixed with a generous amount of ice (an American obsession) and tap or carbonated water. Although I drink and enjoy beer, I consider it to be less healthy for me precisely because of its high caloric content. Most of my weight-conscious middle-class and middle-aged colleagues eschew beer for the same reason. (A glance through an American cookbook for a calorie chart—which I was sure would be there and surprised to see so early in the text—provided the following caloric information: Beer, 12 oz bottle, 173 calories; gin, 1 oz, 70 calories; Scotch, 1 oz, 73 calories.) Evidence that Americans tend not to think of alcohol as food is also suggested by the fact that a night of social drinking is frequently concluded by sharing a light meal or snack provided by a hostess or taken at a restaurant or café. Drinkers of African beer reported that their drinking reduced their appetite for "other" food and thus was sometimes substituted for a meal, especially in the evening.

Some Africans stated a preference familiar to middle-class Whites of eating before embarking upon an evening of social drinking. I suspect that the motives for observing this practice differed in that the African was cutting down his capacity for beer, while the European was consciously ameliorating the effects of his anticipated drinking, either to reduce the impact of what he would consume or to increase his capacity. One African contrasted his own moderate drinking with that of some of his friends on the basis that he never went to the beer garden without eating something first. "Once a man starts to drink, he has no appetite for 'natural' food," he explained.

Although I have described differences between African and European drinking, it should be recognized that the drinking patterns of the latter exerted a strong acculturative influence as models of urbanity. This influence was subtly reinforced by members of the White community through references to the "growing sophistication" of the urban African. Municipal officials in Salisbury and Bulawayo had attempted to accommodate for this sophistication by pushing the development of bar-lounges catering to their more cosmopolitan patrons. Salisbury was devoting a major effort toward satisfying what it perceived (or wanted to perceive) in changing patterns of preference among its customers. Officials in Bulawayo seemed to see their liquor outlets as serving more varied roles, and they invested their efforts and resources in keeping each beer garden functional and up-to-date rather than in developing long-range plans for eventually converting them to bar-lounges. Their efforts to cater to, and capitalize on, the changing preferences of a minority of their patrons included making European beer available at all beer gardens and developing special areas within each beer garden where patrons could sit at tables and chairs and have table service for packaged African beer and bottled beers. Thus they were creating more sophisticated beer gardens, offering options so that customers could sort themselves out according to their preferences and pocketbooks while still retaining their identity as beer-garden patrons. Official policy was to avoid creating or fostering the impression of a status difference between drinkers of European and African beer. To a certain extent this status distinction was precisely the impression that had been created, since European beer clearly dominated the "higher class" establishments, but officials had been successful in resisting the temptation to create exclusive enclaves for drinking European beer within the beer gardens.

Beer Gardens as Loci of Cultural Activity among Urban Africans

The municipally operated African beer gardens of southern Africa represent a successful effort in cultural innovation. In Bulawayo they occupied an established place in the spatial arrangement of the townships and an integral part of the social life. To claim that they formed an integral part of township life does not imply that they were only, and in every way, integrative for the African community; some disintegrative functions will be identified below. Nor does the fact that beer gardens were an acknowledged part of township life mean that they were perforce acceptable to all individuals, any more than a number of other successful innovations (cigarette smoking, cinema, television, rock bands and western-style dancing, shebeens, putting all the rooms of one's "home" under one roof, attending public schools, cooking with electricity) or persistent traditions (polygamy, belief in witchcraft, measuring one's wealth by the number of cattle possessed, paying lobolo, acceding to the authority of the elders) were necessarily accepted by or acceptable to all. Nonetheless, the city's "traditional" beer and the beer gardens in which it was sold were part of the shared body of customs for believing and acting that comprised the totality of urban African culture. The beer gardens will now be examined in one of their most functional aspects—their integrative role in the African community as a place of cultural exchange for those who frequented them in person as well as for others whose lives were affected by beer-garden transactions although they did not themselves enter them.

A Forum

One of the essential functions which beer gardens performed was to provide a forum for the exchange of information. For several reasons, urban life interferes with the lateral transmission of cultural information dependent upon personal networks. First, the kind of information one needs is often held by strangers rather than kinsmen, especially regarding such important topics as available jobs, available women or available recourse in crises occurring when one encounters trouble in a strange setting. Second, information channels are disrupted among urban Africans by a cacophony of language—the chances are that the people next door do not speak one's own language, or at least are not native speakers of it. Within the crowded quarters of the workers' hostels this confounding of language meant that a man might not be able to converse readily even with his most proximate fellows, and so minor an event as selecting among the native-language broadcasts on a radio could produce an element of serious discord. Third, people who did share one's language, special customs and place of origin were present in the townships, but their time of arrival, their varied economic circumstances (for example, whether they resided in a rental area or could afford to live in a township which allowed them to own a house) and other factors tended to randomize their whereabouts. Thus the geographic disruption of residents within the townships imposed a barrier to communication which could only be overcome by a conscious attempt to seek out and meet with persons of shared origin. In this regard, strangers to the city, and particularly those from far-off places like Malawi or Mozambique, were sometimes known to wander about the beer halls listening for familiar dialects as a way of establishing and maintaining contact with their countrymen. People unknown to each other "back in the reserves" became involved in new communication nets in the city. Through such networks they not only established themselves more securely in the new setting but also maintained information links with those who remained at home. New arrivals were sources of news; established city dwellers contemplating a visit back home fulfilled requests to visit relatives, transmit funds, convey messages, and so forth. Each of us lives in such interpersonal networks, typically taking them for granted and seeing them only from his own unique and highly personal perspective. The effectiveness and vastness of such networks among other peoples, particularly those who lack easy access to the technological capabilities of telecommunications, is often a source of surprise, earning such titles as the "bush telegraph" in Africa or Alaska and the "moccasin telegraph" among North American Indians. The beer gardens were not the source of such networks, but for the urban African they served as a living switchboard through which he could avail himself of existing lines of communication.

An Arena

Complementing the opportunity they provided for the exchange of sought-after information and the interaction of persons of similar background or circumstances, the beer gardens also provided an arena in which an individual could expand or sustain his involvement in the potential of new experiences and new contacts made possible by an urban environment. The beer garden represented one setting in the urban complex where an individual could exercise some control over the extent of his involvement with Africans of other cultures and with certain of the accoutrements of Westernized society. He was relatively free to exercise a personal preference for observing and participating in tribal or modern dancing and music-making, he could sit and talk, or he could sit and listen. He could expand his contacts only among others of shared background or he could make acquaintances among those whose language and customs were deemed different. He could stay with the men or flirt with the women. He could participate in the passing parade or remain a passive observer in the manner of many juveniles and younger patrons who reported going to the beer gardens expressly as spectators to "watch the dancing, the singing, and the fights."

The beer garden presented a sort of cultural cafeteria and, unlike constrictions imposed by conditions of housing or employment, each person was free to select from its varied fare. If an individual wanted to overcome social reserve and personal inhibition in order to meet new people, he had the atmosphere of the beer garden to support him socially and the facilitation of drink to help him psychologically. If his predisposition called for restraint, he found his company in a group of like-minded beer-peers. In our interviews with beer-garden patrons we found formally and informally organized groups who met regularly together and shared common bonds in one or more of each of the dimensions on which we probed: similarities in age, sex, and drinking preferences; common kin ties; working for the same employer, holding similar jobs, or being in similar occupational circumstances as job seekers; living in the same neighborhood; coming from the same rural district; and speaking the same language. But we also found, albeit less frequently, groups of drinkers who did not share any of these characteristics, including even one among the trio of most frequently prevailing similarities: all-male associates of similar age who shared a common mother tongue.

The Beer Gardens' "Life of Their Own"

In an earlier section I drew attention to several types of individuals whose social round brought them to the beer gardens in the course of work or the pursuit of leisure. That description emphasized the diverse natures and motivations among categories of individuals, and pointed out ways in which they maintained their special identities in the beer-garden setting. Such mechanisms included the marking out of territory among groups who met regularly and the tendency toward age-grading in interacting groups. There were other distinctions characteristic of special groups, as, for example, the inclination of European officials visiting in the course of duty to remain standing and to ignore the customary greeting and handshaking that otherwise was such an obvious part of beer-garden protocol among Africans themselves, or the reported tendency of some employees in the serveries to become surly toward their patrons as the assembly-line nature of their dispensing job overshadowed the affability usualy associated with a publican.

Beer-garden life was greatly influenced by the cultural milieu external to it and by various subcultures loosely affiliated within its walls. But beer gardens, as one African observer explained, also had "a life of their own." The customs observed there extended beyond the drink itself and the customary way of consuming it to include a set of widely shared beliefs and behavior.

I was limited as an "outsider" to beer-garden culture by reason of race, language and national origin (to name a few of the more obvious differences), and I can provide only a partial catalog of readily observed aspects of this cultural system rather than present a complete picture. At the same time, I do not mean to capitalize on the elusive nature of the concept of culture or imply that a "perfect" ethnography of the beer gardens would explain (or predict) the totality of behavior observable in them. It may be more accurate to think of a beer-garden assemblage as an aggregate of humans who shared most of their traditions outside its walls (e.g., in the case of the patrons, sharing the heritage of the Bantu-speaking people, residing at least temporarily in the townships) and who shared a few additional customs once inside them, rather than looking for a comprehensive and rigid set of behaviors which an individual donned like an invisible mantle before entering the beer garden's busy portals. Here are several facets of beer-garden custom and culture as observed in Bulawayo:

No opprobrium was attached to being in a beer garden without sufficient funds to purchase a beer. White officials reported that the phrase, "Shilling please, boss?" was directed at them as they passed through on tours of inspection and that they were frequently asked for cigarettes. Bar managers suggested that "cadging" drinks was not customary among African patrons; such behavior was considered antisocial under most conditions. One could look for friends or sit alone in hopes of being offered beer, but one was not expected to make a blatant show of begging. Employees and affluent patrons were sometimes put upon for loans, however.

Drunkenness was not categorically denounced, particularly among beer drinkers themselves. Assistance was customarily offered to see that someone drunk got safely home if he had been drinking with comrades. Almost to the man, interviewees who said they did get drunk at the beer gardens reported that they then proceeded directly home and went to sleep. Regardless of whether their actual behavior invariably followed this pattern, it was clearly the accepted social norm. In terms of assessing drunkenness, one African observer noted, "If a man has been drinking and he stands up and announces, 'I'm drunk,' then he probably isn't. But if he argues that he is not drunk, then we know that he is." The thickness of the beer and its relatively low alcohol content probably added to some ambivalence in distinguishing whether one was actually inebriated or merely sated. To our standard interview question, "When was the last time you got drunk?" we occasionally received an unanticipated response: "I'm drunk now." Men who frequented the beer gardens in search of female companions, and who were expected to evidence their interest through purchases of beer, reported with some resentment how they were often led on by women who appeared tantalizingly inebriated in one group but, if the money ran out, suddenly appeared quite sober when joining another man or group of men.

Our interviewees generally described themselves as peaceful and happy drinkers who did not approve of displays of aggression by other drinkers and who did not engage in such behavior themselves. Drinkers as well as those who held some responsibility for looking after them (e.g., bar managers, policemen) drew a line between the drunken behavior of "normal" drinkers and troublemakers. The former posed no threat to anyone else. For the most part the "normal" drinkers described themselves as becoming either more introspective or more amiable, the latter type inclined to entertain their friends with jokes and stories, to join in singing or dancing, or in other ways to exhibit their euphoria. Although they might stagger a bit, they were able to get safely home and to bed.

At a more complete state of intoxication were the helpless "drunks," typically older men whose drinking was heavy and less social in nature and who became so insensate that they appeared not to know what they were doing or where they were. Except for their not-particularly-appreciated habit of uninhibited urinating anywhere they happened to be, such persons posed no danger and little bother to others. Once they had become "sloshed" they usually fell asleep in the beer garden or along the attempted route home.[8] In such a condition of drunkenness they did pose a danger to themselves, however, since they were easy victims of thug attacks (sometimes by youthful gangs whose only motives seemed to be to inflict terror on others) and accidents.

The aggressive "drunks," women as well as men, became raucous and offensive, shouting and swearing at others and becoming in-

[8]Cashiers were instructed not to continue selling beer to persons who appeared drunk, a policy that afforded little control under crowded and busy beer-garden conditions. Such a policy was a reality in the more exclusive setting of the bar-lounges. As one White employee of the municipality explained while describing bar-lounge standards to me, "You have to wear a tie or cravat to get into those places after 6 PM and if anybody gets pissed [drunk] they chuck 'em out!"

creasingly quarrelsome, disruptive and uncontrollable. Such behavior violated prevailing norms. Blame was sometimes attributed to the liquor but was most often leveled at the drinker himself; one was not excused for misbehavior while drunk.[9] As one interviewee noted, "It goes according to how you are—it's not the beer. If you are short tempered, it's worse with beer."

Another man explained, "In most cases people will try to blame the beer, saying, 'I was drunk, it was the influence of beer.' But it is a man's own choice to drink. He assumes responsibility. Beer doesn't move. It doesn't do any talking. You can't blame it. You are the master."

In terms of more serious cases of aggression in the public setting of the beer gardens, observers noted that aggressors usually knew full well what they were doing, were never too drunk to fight, and sometimes came to the beer gardens precisely because they offered such good places to start an altercation. One bar manager pointed out, "The ones who make violence are not the ones who are drunk."

Although fighting was not "acceptable" it was nonetheless an accepted part of beer-garden behavior and an invariably mentioned aspect of beer-garden lore. Administrators, employees and patrons expressed concern over the fighting and recommended that more policemen be stationed in the gardens for security and protection. Officials of the Housing and Amenities Department expressed special concern because reports of assaults made in beer gardens or otherwise linked to urban drinking reflected on the effectiveness of their administration and increased their vulnerability to outbursts of indignation from vigilant monitors in both the African and the European communities. Wooden knobkerries and knives were the most commonly used weapons in assaults reported in the townships.[10] Officials were endeavoring to have the Okapi knife, a weapon which apparently had replaced the previously banned "flick knife,"

[9]This is in contrast to the pattern more frequently observed in my fieldwork among the Kwakiutl, where attempts were regularly made by a person who had been on a binge and by kinsmen acting on his behalf to lay the blame on liquor rather than on the individual who had consumed it. "He couldn't help it, he was drunk," was a frequently-offered explanation for drunken behavior (138, *p.* 57). Similar explanations in the African setting appeared not to be regarded as convincing.

[10]According to an account in the local newspaper, a knife was used in each of the 4 murders and 73 reported cases of assault occurring in the townships during October and November 1970 (Bulawayo *Chronicle*, 17 December 1970).

declared illegal. "I don't know why, but people in Bulawayo are said to be 'trigger-happy' with their knives," an African administrator explained.

Knives were the most lethal among the instruments identified in serious cases of beer-garden assault, but other handy weapons were also enlisted on occasion. Metal beer mugs that had once proven popular had to be retired as they came increasingly to be regarded as weapons. Stones picked up from the ground were also used, smaller stones of up to 4 pounds being thrown, sometimes in a mob stoning, and larger ones being used in attempts at skull crushing. Stoning within beer gardens was deemed to be a sufficiently persistent danger to warrant a standing order holding managers responsible for seeing that stones were never left lying around loose inside the gardens. This order could in fact be observed only at improved sites, and a rash of improvements being made during 1970–71 left stockpiles of bricks and other construction materials handy as potential missiles in several gardens.

Characteristic of many African tribes in general and perhaps of the Ndebele in particular, the air of suspicion that pervaded much of everyday life (24, 122) was evident in beer-garden life as well. Beer gardens provided a fertile environment for rumors and distortions of fact. Sometimes the beer itself was the focus of attention, with accusations that the beer had been adulterated, that it was intended as a political pacifier, or that it contained substances for effecting birth control. Such stories are reminiscent of those circulated among American servicemen that their food contained dosages of saltpeter to reduce their sexual drives. In the case of the municipal beer, it was fashionable to blame the beer for diminished sexual interest: "You come from the beer garden late, thinking only of sleep." However, beer was also held responsible for a general loosening of morals, and thus was blamed for cases of sexual promiscuity and pregnancy. When the beer itself was not under question, the beer garden nonetheless functioned as a breeding ground for rumors of more far-reaching consequences. The recurrence of real and imagined problems with beer quality, the real and imagined omnipresence of uniformed and plainclothes policemen and the constant flow of accurate and inaccurate information provided ample fuel for maintaining the aura of suspicion.

My own interest in studying beer gardens raised some apprehension about how such inquiry might be perceived by patrons. An

African official cautioned, "I think we may have overdone our talking about beer drinking. The very fact that we were interested in it in conducting our household survey raised questions on the part of African people. I definitely would not be a party to going into the beer halls to do any interviewing."

Two African assistants assigned by the Housing and Amenities Department to aid me in interviewing beer-garden patrons spent days helping me develop and pretest an interview schedule and listening to my plans for randomizing our selection of persons to be interviewed. Their appreciation of the need for randomizing interviewees evaporated in the face of far more pragmatic considerations the instant we first set foot in the beer garden where we planned to conduct the research. It would be the very randomness of their selections that would most arouse suspicion, they explained. If they were to manifest such strange behavior only once or twice, perhaps no one would notice, but the fact that we were going to return to the same beer garden day after day for a period of 3 weeks could easily create suspicion resulting in anger and perhaps even bodily harm directed toward them. Similarly, White officials who did not seem opposed to having the interviews conducted nevertheless cautioned that they should be carried out in the semiprivacy of the servery where they would attract less attention and "keep everyone from getting too excited." My assistants proceeded very slowly in extending their first invitations for interviews, always keeping in mind how they might be received in ensuing days. What they had not anticipated was that our returning day after day would eventually begin to work in our favor as word spread of what we were doing (we described the study exactly as it was, "To find out what kinds of people use the beer gardens and what they think of them"), and we found people whom we had interviewed coming by to greet us or to introduce others for us to interview.[11]

[11]Although we were unsuccessful in randomizing our sample on a counting dimension, we varied it as we could along other dimensions by varying the hours and days when we interviewed and by letting other people select interviewees for us rather than select them all ourselves. We included interview periods to cover daytime and evening drinkers during the week and the weekends. We still could not be sure what our sample was representative of, so we turned our attention back to the total population of persons present in the beer garden. We conducted headcounts by sex and estimated age during periods when we were interviewing, since it was quite obvious that women were grossly underrepre-

Beer gardens had a potentially volatile atmosphere, particularly during crowded weekends in periods of hot weather. An African observer summarized it succinctly: "Beer-garden people can be quite emotional." Among the routine precautions for maintaining control, managers were reminded to summon outside help at the least indication that a random crowd might coalesce into a hostile mob. Old-timers in the Department readily recalled the severity of rioting in the early 1960s when some beer-garden staffs remained locked in central serveries or offices, crouched under the furniture to withstand a pummeling of rocks while they awaited rescue from mobilized forces. Although Europeans and Africans alike were optimistic that times had changed and a recurrence of such riots was unlikely, beer-garden violence remained both a threat and a reality.

Most violence was interpersonal, resulting in fights and stabbings, but incidents of collective violence and mob reaction did occur. If the stories recounting incidents were not necessarily accurate they nonetheless served effectively in keeping everyone mindful of how quickly a beer-garden crowd could get out of hand. One such incident had occurred in 1970 when a series of strange and essentially unrelated events was triggered off, the origins of which were traced to a casual and well-intentioned act of an African undertaker. The ensuing set of events was predicated on the rumor that the undertaking business had been poor and so local undertakers had set about poisoning municipal beer. In each beer garden affected, the sequence of events developed somewhat independently, but the end result was that during one weekend many people left the gardens to escape the possibility of being poisoned or being accused of attempting to poison others. Sales of African beer were severely affected. Violence erupted and several people, variously reported numbering from one to seven, were left dead in the wake of the pandemonium. The extreme variation in accounts given by Europeans and Africans alike suggested the impact of the story even in the absence of accurate facts and attested to the awesome fury residing in the hands of beer-garden patrons. Officials noted with a partial sense of relief that it was not municipal beer itself that was

sented in our sample and we were ignoring under-age juveniles and children completely. At first my assistants were also reluctant to conduct headcounts at the gates, since a headcount showing sex and age required a visible tally form rather than the use of simple and easily concealed hand counters used in earlier surveys conducted by the Department.

held accountable as the source of the trouble in this case, as had often happened before.

Patrons exhibited strong loyalties and preferences for particular kinds of beer, particular serveries, particular managers and employees, and particular beer gardens, often to the frustration of attempts on behalf of officials to improve facilities. At the Main Beer Hall, for example, some customers waited patiently until a later-opening unit began serving in order to buy beer at the servery operated by a preferred manager or to obtain beer served through a favored Buck dispenser. Each servery at Main Beer Hall and at the larger gardens had more than one dispensing unit, but every garden had a "favorite side" in terms of customer preferences. Long queues might form in front of that side while employees stationed at the alternate meter, dispensing exactly the same beer from the same tank and connected by clearly visible systems of pipes, stood idle. Daily sales patterns showed volume sales of several times the difference between the more and less popular sides of the same servery.

Patrons also exhibited preferences among the containers available for them. Newer mugs were selected in preference to older ones, and large gallon mugs were selected in preference to half-gallon ones. There was a popular notion that "beer always tastes better out of a 2-shilling mug." In Salisbury, where the majority of mugs were of half-gallon size, no comparable preference was reported. However, a choice existed there between two available colors of mugs: white and dark blue. Barmen noted that customers buying gallons of beer were not particular about the color of the mug, but customers purchasing only half gallons preferred the dark mug. Since one could not see the amount of beer in the dark mugs, it was assumed that persons who purchased the small amount of beer out of economic necessity preferred a mug that did not readily reveal the quantity of beer in it. Possibly a comparable self-consciousness in the "presentation of self" was the underlying factor in the tendency of Bulawayo drinkers to select a gallon container even if they only intended to purchase half a gallon of beer.

If a new batch of beer differed too markedly from the beer served earlier in the day, customers reportedly had been known to leave a beer garden en masse and go to another garden nearby where the taste of the beer was judged to be more satisfactory. "Word that the beer is 'off' at some beer garden gets around very quickly," one supervisor explained. Closely paralleling some of the worst accusations made about the beer, patrons also circulated stories attesting to the especially desirable qualities of whichever product was their favorite. Each of the five bottled European beers available in Rhodesia, all brewed by the same parent company—"Black Label," "Castle (Lager)," "Lion (Lager)," "Milk Stout," and "Pilsener," had its adherents, but Castle Lager was far and away the favorite. Individual references were made to brands "too strong" or "too weak," although European beers were not thought to exhibit as wide a range of properties or as great a variation from batch to batch as African beer. One of the particular attractions of European beers was that they could not be tampered with, diluted or added to before they were opened, an attribute that made them particularly attractive under any circumstances where one might question the integrity of a private host (e.g., a shebeen operator) or a publican. Whether drinking European or African beer, patrons usually preferred that it not be served too cold. The stated reason was that cold beer was "slow to take effect" and a person might not feel any immediate effect from his drinking. Later, as the beer warmed up and its potency heightened, he might get very drunk while asleep and not know it. Worse yet, he might "pass out or even die" from having taken too much beer and never realizing it at the time.

The bulk handling of African beer rendered it easily susceptible to accusations and feelings of suspicion, but the beer enjoyed a wide spectrum of desirable qualities attributed to it as well. There was, for example, a general feeling that Bulawayo's beer excelled Salisbury's, and officials reveled in the praises sung to their beer by local Africans. Along with this feeling was a notion that African women in other parts of Rhodesia admired the virility of Bulawayo men, a quality that was somehow linked with the beer they drank. Bulawayo's draught beer, Ingwebu, was literally millions of gallons ahead of the sales of the packaged Indlovu and Jabulisa, although they all came out of the same vats. The higher-priced packaged beers were, however, marketed as "select" brews put up during the slack part of a week when there was adequate time to let them ferment, and thus there was some basis for the popular belief that packaged beer was stronger. As noted, some drinkers laced their draught beer with the addition of a quart of packaged beer. It was difficult to explain the preferences stated for one or the other of the beers sold in the two differently labeled packages and to account for the far greater market which Indlovu enjoyed over its identical twin. Indlovu had grown up the favorite; Jabulisa had earned an unfortunate reputation for eroding the prowess of men who drank it, and women were said to chide husbands who showed no interest in sex after spending an evening drinking for having "drunk too much Jabulisa." I once heard the suggestion that husbands who could not perform, particularly if their lack of interest was due to an amorous adventure earlier in the evening, sometimes found the same story useful in explaining their disinterest to their wives.

Summary

Given the objective of presenting urban African drinking in Bulawayo as a study of culturally integrated drinking, the major portion of the brief for the case has been presented in this chapter, and its dominant features can now be summarized. I have identified several aspects of drinking which comprise the formal category "Integrated Drinking" as defined in studies of cross-cultural comparisons of drinking practices or which are typically found in association with such drinking. These include ceremonial use of alcohol, ritualization in alcohol use, an ample rate of consumption, acceptance of drunkenness, and relatively frequent drunkenness. Admittedly several of these categories are rather subjectively evaluated, but it can be pointed out that many urban Africans describe their own drinking behavior and drinking attitudes in a manner consistent with these descriptive categories.

Patterns of behavior regarding alcohol use date from precontact times, and the "traditional" beers brewed by the municipalities and commercial breweries provide a continuity between past and present and between rural and urban life styles. The "new" brewers have capitalized on the African attitude that beer is a food and have stressed this aspect in expanding their markets. Offering Western-style beer as an alternative to African beer has served to introduce change via an elective category of behavior—the young man wishing to identify with his new life in the city can sip his bottle of European beer or share a mug of traditional brew, as his personal preference moves him and his pocketbook allows.

The creation of formally designated beer gardens in urban centers has tended to institutionalize drinking and to focus attention on a single aspect of traditional life, creating an atmosphere of a drinking den more characteristic of Western society than of the indigenous one. It is interesting to consider how urban Africans have imposed their own meanings on the gardens. Instead of regarding them as "clubs" or "bars," they have assigned them a much more integral role in city life. As community centers providing a focus for social networks, the beer gardens function in the manner that other types of drinking establishment do elsewhere in Africa.[12] The beer gardens serve tribal affiliations as well as nontribal ones, functioning as an exchange where persons dissociated from the people back home can make or renew contact. Each individual could use the beer garden to ameliorate the dissonance of his own urban–rural discontinuity by maximizing or minimizing the extent of his nontribal network.

The gardens serve as a place to find work as well as a place to find leisure, a place to spend time or to "kill" time. Indeed, enthusiastic patrons and perennial critics alike agreed that the beer gardens were the action centers of the community, and that no other facility had a comparable vitality or attraction, although it is necessary to point out that advocates of other activity and interest patterns (e.g., sports, church) argued on behalf of their commitments with equal vigor. One enthusiastic architecture student completing an academic project in beer-hall design at a university in South Africa presented an interesting alternative to the problem of separating beer garden and community by suggesting that the beer garden become a more integral part of the community rather than a discrete unit within it: "It will be my architectural solution to bring life into the beer hall and for the beer hall to reciprocate and impart some of its life into the community. It will no longer be an isolated unit—a beer and liquor 'outlet,' but will find its appropriate place within the community it is to serve" (3, *p. 32*).

Perhaps the beer gardens provided the African in the urban setting a down payment on the promise of the city, a compensation for the growing sense of isolation associated with modern industrial societies and the accompanying anxieties and frustrations of putting one's fate in the hands of strangers. True, for too many the down payment of happy hours spent with colleagues in the beer gardens may have been one of the few rewards the city had to offer, but at least it provided a setting where one could revel in the freedom of urban autonomy and the euphoria of temporarily forgetting his woes.

In the next part I shall expand the analysis of integrated drinking

[12]See Gutkind's account of the beer bars in Kampala (44, *p. 394*).

beyond the boundaries of one culture to explore how urban African drinking involved members of the European society in a variety of professional roles and thus also served some integrating functions between cultures. I shall also present the views of many Africans in direct contact with the beer gardens or otherwise involved with urban African life because of professional roles they occupied. The behavior and attitudes of urban Africans in their personal roles as residents of the township will be the focus of part IV.

Part III

Professional Involvement with Urban African Beer Drinking

5

Role Dimensions

THE PICTURE of urban African drinking emerging here has been painted with a broad brush. Constant reference has been made to collectivities of people, like beer drinkers, officials, Africans and Europeans, representing these groups as idealized types and largely ignoring within-group variation in adaptations to the municipal beer gardens by the African population who were their intended clientele. Here urban African drinking will be examined from the perspective of Africans and Europeans whose formal involvement with the process was related to their occupational role.

How persons who occupied the roles to be discussed here coped with drinking in their private lives was a personal matter influenced by the formal roles they occupied but not necessarily circumscribed by those roles. The fact that the municipal beer gardens catered only to Africans automatically excluded Whites from informal social participation in beer-garden life even while it allowed, or even required, the presence of some of them in enacting their formal roles. It should also be pointed out that although the formal roles described here could be and in fact were occupied by both Whites and Blacks, the roles themselves were entirely Western in conception and Africans occupying them were highly acculturated. The ultimate authority for each role remained in White hands. Churchmen appeared to be the most self-conscious in this regard and had made the greatest display of effort to "Africanize" the ranks of their hierarchies. There was a well-entrenched belief among a majority of White professionals in every field that the African was not yet ready for positions of ultimate authority and responsibility, and a selective perception of the problems of countries to their immediate north provided constant assurance to White Rhodesians of their wisdom in this matter.

The problems and points of view will be presented here as stated in formal interviews and as observed throughout the course of the fieldwork. It may be well to remember that the anthropologist's interest in human behavior leads him to a pervading interest in reality as individuals in different cultural settings perceive it. Views quoted throughout these pages are not offered because they are right or wrong but because they are statements about what people thought, what they said and what they believed they had observed. The section on physicians incorporates an exception in that an attempt has been made to provide some generally accepted scientific evidence by including references to relevant medical research in addition to presenting the perceptions of practicing doctors.

The professional points of view represented include, in order, those of churchmen, welfare workers, police, physicians and administrative personnel most directly associated with the Liquor Undertaking. In an earlier chapter reference was made to beer-garden employees—as well as a collectivity of others for whom beer-garden patrons provided a source of customers or income—as "beneficiaries" of the beer gardens. I think it is instructive to consider all the persons discussed in the present part as beneficiaries of the beer gardens, not only in the immediate sense that the source of funds for the salaries of some of them (e.g., welfare workers and certain other employees in the Housing and Amenities Department) were derived directly from beer profits, but in the sense that all of these professional roles were defined in part by the presence of drinkers and conditions relating to drinking. Even churchmen, whose characterization as beneficiaries of beer gardens may seem outrageous, defined part of their role in terms of the stand they took and the behavior they exhibited toward drinkers and drinking, regardless of whether they participated in any way themselves. Thus at least implicitly each role is to some extent integrated because drinking provides a focus—or foil—for that position. Were drinking suddenly and totally to disappear from Bulawayo, or Rhodesia, or the whole of the "Dark Continent," occupants of each of these professional roles (and many others that will not be discussed) would find their own set of behaviors and expectations having to change, some (like bar managers) radically, some (like clergymen in churches not rigidly opposed to drinking) only tangentially. I shall return to this notion of drinking as an integrating factor within professional roles, among them and between European and African society, at the conclusion of this section.

6

Professional Points of View: Churchmen

INDIVIDUALLY AND COLLECTIVELY the Europeans with whom Africans have been in contact for generations have represented a potpourri of opinion and example about drinking among Whites, from hard-drinking adventurers to adamant abstainers. (Conversely, the African seems to have provided his acculturators with a range of attitudes and behavior toward drinking, although he has been less compulsive about proselytizing for one view or the other.) Although many churchmen have been quick to condemn the disintegrative aspects of imbibing, collectively they have confronted the African with a marvelously disparate set of opinions and beliefs concerning man's proper relation to the liquid spirits. Contemporary churchmen in Rhodesia lined up on an expected continuum. Roman Catholics took a liberal position, leaving the use of alcoholic beverages to individual conscience and concerning themselves only with curtailing excessive drinking. "Catholic missionaries have never told people they must not drink," explained a priest and long-time missionary to Africa. "We are not prepared to say it is forbidden by Divine Law." Protestant sects held positions ranging from attitudes tolerant of moderate drinking to intolerance of any drinking at all. The indigenous apostolic churches took their place among the strictest in the latter category (89).

What an American Missionary Said About Drinking

The following excerpts are from a tape-recorded interview held in Bulawayo with a Protestant missionary. This man, a person of long experience in church work in southern Africa, described his own position on drinking as liberal.

"The problem with the beer drinking is probably the preeminent social problem in the country today.

"I don't make the statement that a man should never touch alcohol. That's not the basis of my argument. *Temperance* is the lesson of Scripture, not abstinence. But the people rarely drink this stuff in temperance. They drink to excess.

"In traditional African life, beer was strictly controlled. No one except the old men were supposed to drink it. That is not to say that there were no violations, but there were sanctions on it. In the urban society, there are no effective controls on the use of beer. They say that children are not supposed to go in the beer halls, but I've seen children near the gates and hanging around inside. My African friends and I have seen beer handed over the walls to children who were supposed to take it home but who in fact drank it. Beer can be bought by an older person and given to the younger children outside, even though there are statutes regulating the age at which drinking occurs. There are no controls regulating the age at which drinking occurs. There are no controls over the men drinking, or over men and women drinking together, and there are no controls on when they can drink. The beer halls are always there, serving as a constant invitation to them to drink.

"The African viewpoint, and I've heard it repeatedly, even from drinkers, is: This is a deliberate plot to keep the people down. The government is trying to encourage them to drink. If the people drink up their money and strength, they can't rise up. The beer halls sap their vitality. It's a pacifier, by the African's own admission.

"It's not a deliberate ploy or effort; but the African thinking is, 'The White man is so clever and smart, and he sees that we have a weakness, so he makes the beer available. That's an awfully good way to keep us down.' Why do the Africans resent the beer halls and drink at the same time? I think people would generally say, 'I was tempted.' If the beer is here, it is here to be drunk. The blame is on the temptation. Similarly, if a girl comes along and makes a proposition and they go and sleep with her, it's not their fault. The internal self-discipline is not there. This ties in with the group-centeredness of African life, an unwillingness to rise above the mean. It's all part of the same psychology. They do what is there to be done.

"The churches do not need to take overt action against the beer halls themselves. Our group is not going to start a vendetta against them. But we should work for the introduction of more social controls. The Europeans have broken down the old social controls without generating any substitute ones. I would really be amazed if any municipality would reinstate the traditional controls on beer drinking, since this runs counter to their motives to make a profit, but I think this would solve a great many of the problems associated with beer halls. For example, if they would reduce the size of the beer halls, have closer supervision, reinstate segregated drinking, or revert to the traditional patterns in which drinking was limited to certain statuses and times, or was to celebrate the completion of certain jobs, as a reward rather than as a pacifier.

"I've heard the argument put forth by many municipalities—not just

Bulawayo—that it's better that we produce this low-powered stuff that does have certain nutritional value and is brewed under hygienic conditions and try weaning them away from high-powered hard liquor that they get either by purchasing European liquor, making their own, or going to the shebeens. Obviously, there's some degree of truth in that. But there is enough alcoholic content in this beer that they show every symptom of being drunk and disorderly, with a loosening of morals and of all restraint. Every bit of behavior associated with the use of alcohol and intemperance is associated with the beer halls.

"Maybe the officials don't actually know what happens in the beer halls. Like any bureaucracy, maybe those in charge don't really know what is going on. Or else they accept this type of behavior as if it's the normal behavior of African people. Beer halls have degraded the quality of African life. I resent the boosting of the beer halls on the part of any agency that should be working to help the people. I don't understand why men as astute as those who head up the Housing and Amenities Department allow it, although the policy here is the same as that you'd find in cities all over Africa.

"Zambia, too, has fantastic problems. I've challenged some African friends there. 'You have always said you drink because of your frustration at a White government. But what about the drinking problem right here in Zambia?'

" 'Well,' they answer, 'the Whites taught us how to drink. We haven't yet had time to shake the habit.' "

What Other Churchmen Said About Drinking

I became particularly interested in the attitudes of clergymen from the more fundamentalist Protestant sects toward African drinking and availed myself of any opportunities that arose during the year to discuss the issue with them.[1] To a man, each of them explained that drinking was not a problem among members of his church. The explanation was simple. Drinkers were not allowed to become members or, in a few churches that tolerated them, drinkers were not accorded the full blessing of the Sacraments. "We would say that a person who drinks is not fully committed to our position," explained a Brethren in Christ missionary.

[1]Specifically these included such organizations as The Evangelical Alliance Mission, African Reformed Church, Pilgrim Wesleyan, Brethren in Christ, Free Methodist, Southern Baptist, Central African Baptist Union, Church of Central African Presbyterian, Church of Christ, and Salvation Army. In Rhodesia, as in other areas with an English heritage, these denominations are referred to as the "nonconformist" churches, i.e., Protestant sects which have broken away from the original Church of England.

"We don't have anything in our church laws against drinking," explained a visiting missionary from Zambia, "but how can one be a Christian if he drinks?"

"If we find that a member of our church has started drinking, we just warn him," summarized a rural African pastor in the Seventh Day Adventist church, "and then if he keeps on, we drop him."

Coping with the problem of drinking by excluding drinkers from full participation in their churches struck me as a rather unusual way to provide help for people (95, 140) although churchmen noted that in fact the persons who turned or returned to drinking drifted away from the churches on their own—they were not formally banned. Pastors and church members hastened to describe their efforts to help individuals overcome their drinking problems to meet the prerequisites for church membership. Considering the role drinking has played in traditional society, I wondered whether the fundamentalist churches were systematically excluding male heads of households from their congregations. Sangree (112) has described a case in Kenya in the mid 1950s where the failure of missionaries to recognize the central role of beer among the Bantu Tiriki resulted in a church membership of persons marginal to the existing culture: uninitiated aliens, witch suspects and outcasts among the senior men, and a preponderance of women and children. My missionary respondents were ambivalent about whether their congregations were composed disproportionately of women and children. Officials of at least some denominations agreed that men were underrepresented; officials of others argued that this was not so. One interested observer offered the broad generalization that "the Catholic church is essentially a church of children, the Methodist a church of women, and the Apostolic a church of men."

Whatever position their churches took toward drinking, clergymen called my attention to their own individual efforts to be "realistic" in their views. Their realism followed two forms, represented in two statements quoted below. The first, as illustrated by an African minister, was to attempt to be sympathetic in their attitude toward infringement:

> "Our church doesn't approve of drinking or intoxication, but you know how people are—when the cat's away the mice will play. We have a regulation that if a church elder sees someone drinking or drunk, he can bring that person before the church council to be warned or to be disciplined [suspended] for some period. It doesn't happen very often. I don't recall

it happening last year. If one goes to the beer garden, who is there to see him?"

The second point of view, represented by a White missionary, was to be liberal in one's personal behavior but to keep up appearances in one's formal role. I do not mean to represent fundamentalist clergy and missionaries as incipient alcoholics, but I was fascinated by the number of Whites—and even a few Africans—whose personal attitude toward imbibing differed from the official doctrines of the churches they represented. Did I perhaps even sense a feeling of relief among some of them when they realized they had a sympathetic ear?

"Actually, my wife and I occasionally drink wine with our dinner, and we go to dances. She wears short dresses and has pretty legs, and I wear my hair long. But of course I'm supposed to be a teetotaler, and I could never let it be known that I drank. We sort of go opposite to a lot of the things that most missionaries do. We think it works better."

Most churchmen cut a wide swath around the beer gardens and claimed little knowledge of what went on inside them. None, including African pastors for whom such a notion might have been a reality, expressed an interest in them as a place for learning about urban life. Apparently perceiving his own role solely as a "giving" one, an African minister explained that there was no point in going into a beer garden unless one had a loudspeaker—which, he added, the city council would never approve. "I don't know of anyone going into beer gardens to visit—it's too rough and rowdy," he explained. But another missionary who said he was sorry to report that not much was being done or planned by his group about drinking, or about offering an alternative to it, observed, "We offer no substitute for the fellowship of the beer garden."

Among more formal lines of communication, some efforts had been made by various Christian presses to warn of the perils of beer drinking. Religious publications sometimes included articles or stories about drinking. A modest little booklet of 28 pages available for 7½¢, *Doro Kudya [Beer is Food]* (87), published in Shona, had enjoyed a printing of 10,000 copies and had subsequently been published in a Ndebele version in 2 editions of 3,000 each.[2] A more ambitious 124-page "novel" in Shona dealing with African drinking

[2]Mambo Press. [Personal communication, 1971.]

PLATE 1.—*Traditional Brewing; Women Preparing Beer in 44-Gallon Petrol Cans.* (From a painting by Matthew Zulu owned by the author.)

PLATE 2.—*Farm Workers Near Salisbury Enjoying a Weekend Beer Drink.* (Photo by the author.)

PLATE 3.–*Drinking Traditional Beer from a Gourd, Matapos Hills Area.* (Photo by the author.)

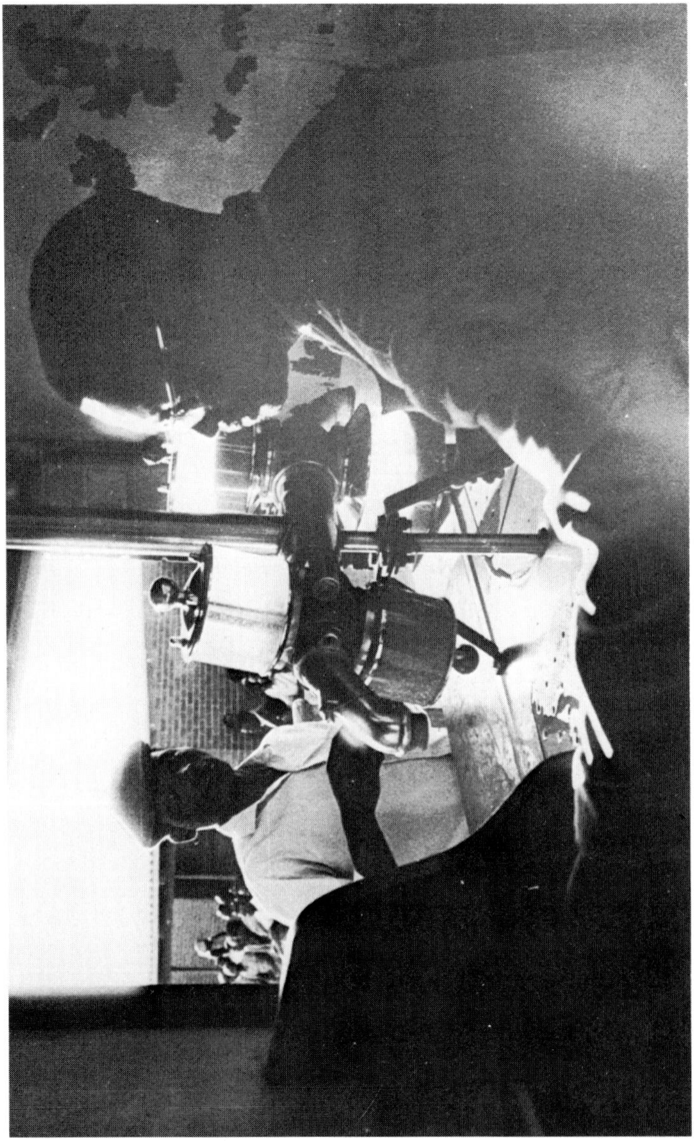

Plate 4.—*A Beer Garden as Seen from Inside the Servery.* While the cashier makes change, the customer's beer mug is filling from the Buck dispenser at the right. (Courtesy of Housing and Amenities Department, Bulawayo.)

PLATE 5.—"*When the Beer Is Good, We Rejoice.*" Afternoon scene in a Bulawayo beer garden. (Courtesy of Housing and Amenities Department, Bulawayo.)

PLATE 6.–*Beer Garden Scene.* (Courtesy of Housing and Amenities Department, Bulawayo.)

PLATE 7.—*Beer Garden Scene.* (Courtesy of Housing and Amenities Department, Bulawayo.)

PLATE 8.—*Members of an African Burial Society after Concluding the Business Portion of Their Meeting.* Beer was provided by fines levied against latecomers. (Photo by the author.)

PLATE 9.—*Boy Drinking Packaged African Beer.* (Photo by the author.)

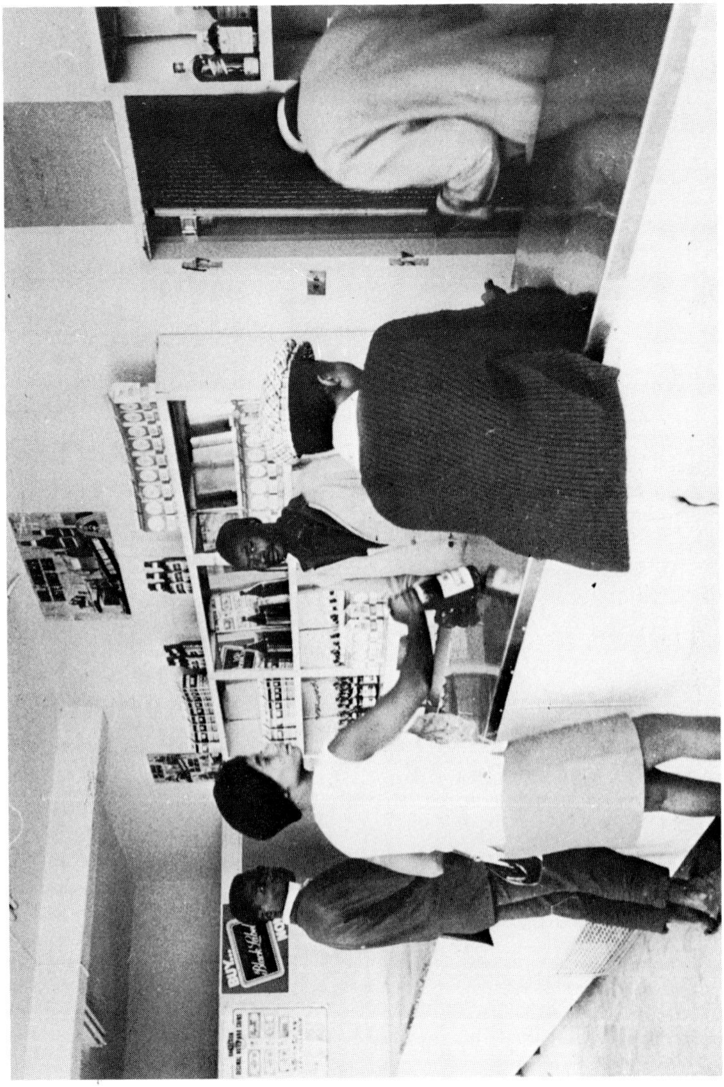

PLATE 10.—*Township Residents at Muncipally Operated Liquor Stores.* (Courtesy of Housing and Amenities Department, Bulawayo.)

PLATE 11.—*A Bar-Lounge Facility, Bulawayo.*
(Courtesy of Housing and Amenities Department, Bulawayo.)

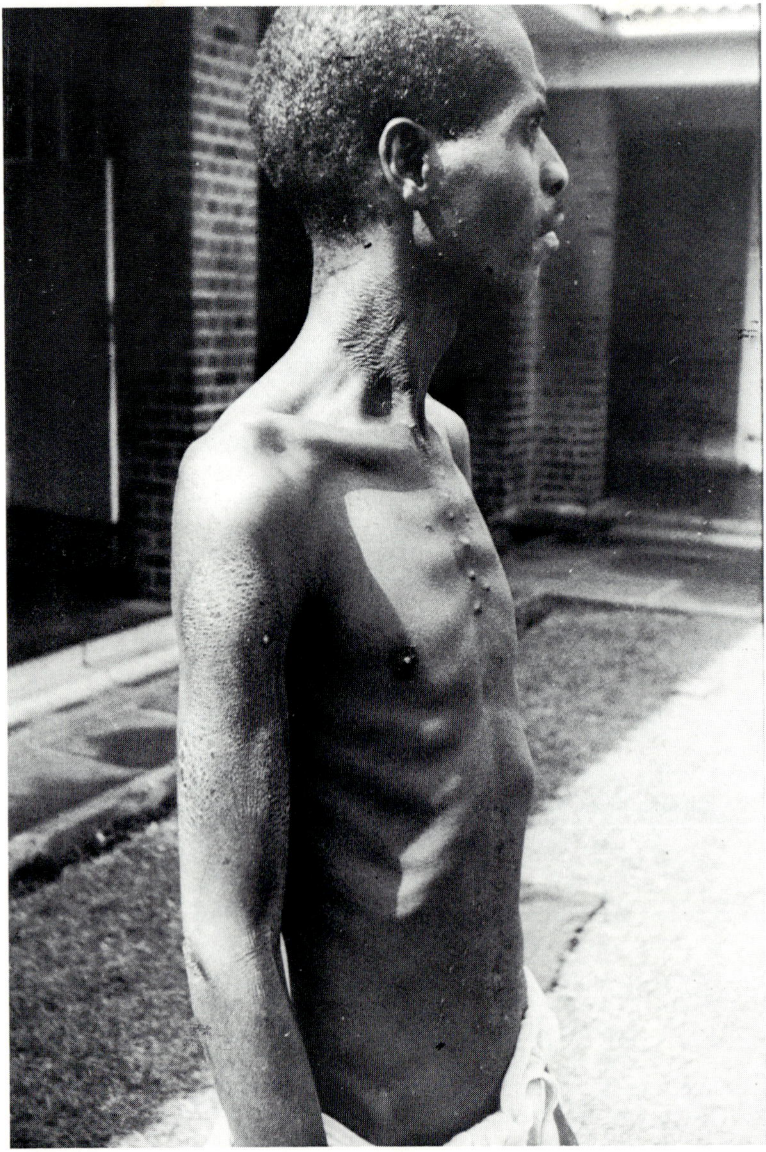

PLATE 12.—*A Case of Pellagra.* Relatives reported that he had "only drunk beer" and eaten no food during the previous year. (Photo by the author.)

(23) was already in its 2d edition. A small portion of the book jacket devoted to English lured the would-be reader or buyer with this message:

> "The Magic pen of Chakaipa paints a stark picture of the dire consequences that befall a family when father takes to drink. Frustration and despair lead the mother down the same slippery slope . . . the children are neglected . . . the marriage breaks down. These evils are interwoven into an enthralling plot which contains all the ingredients of a first class story."

Films depicting the perils of drink and carrying such titles as "None For the Road" or "What About Drinking?" were also shown to African audiences. Films were distributed from central lending libraries and coordinated by a national church council. One 20-minute color production, "I Am Guilty," had been filmed in the Salisbury townships under the direction of an Evangelical Lutheran missionary and included a rather predictable sequence of scenes shot in one of Salisbury's beer gardens. The film was reported to have received wide circulation in the late 1960s.

7

Professional Points of View: Municipal Welfare Workers

PERSONNEL employed in the welfare services of the Department of Housing and Amenities found themselves in a paradoxical situation in regard to the municipal beer industry. The most obvious aspect of the paradox was that the Liquor Undertaking which served as the source of many of their problems was also directly and solely the source of funds for coping with them. For welfare workers to categorically reject the beer industry—as certain critics among churchmen and civic minded townsmen could do—was literally to bite the hand that fed them.

The Department's Senior Welfare Officer was frequently called upon to explain (and sometimes to defend) a welfare program dependent on the municipality's sale of beer. His explanation, such as the following excerpt from a formal paper, always included a reminder that not all the beer profits went into welfare. One-third went to housing and another third went to health services; the latter were also subsidized by the national government.

> "The well-known relationship between welfare and beer profits gives rise to the criticism that we encourage poor people to spend money on liquor, so that we can supply remedial services which they might not need if they drank less. Personally I doubt very much if the availability of a wholesome brew at a reasonable price causes people to spend more than they otherwise would. Attendance at shebeens and house parties, where the tariff is high, suggests that other factors predominate. But you may well feel that it is not desirable that the provision of welfare services should depend upon the profit made from selling liquor. This I would agree with, although I think it a sound principle that such profits should go into the public purse. I must confess, however, that if welfare had to compete in the open market for funds, I could not hope to get as much as I do by having a private fund to draw upon" (34, *p. 2*).

The numerous Africans employed in the Welfare Section (some European case workers were also employed), especially those professionally qualified as case workers, were influenced in their attitudes toward the municipal brewing industry by other subtle factors. Excessive criticism of the organization was not felt to be especially appropriate anywhere within the bureaucracy. Even administrators in high positions maintained a respectful distance between themselves and the Director and expected their White subordinates to do the same in a status hierarchy that worked efficiently and, to lifelong British-trained civil servants, seemed as natural as it was predictable. Africans who had achieved some mark of recognition by those at the top of the hierarchy seemed particularly mindful of differences between statuses achieved (e.g., occupational position) and statuses ascribed by birth (e.g., skin color). They were also watched by their European colleagues for indications of dissatisfaction with the status quo. Reference was made during more than one occasion in discussing Africans who had successfully worked up to a good position to a characteristic "chip on the shoulder" attitude not especially coveted. But just as the Housing and Amenities Department was said to represent about the best employment opportunity for an African in terms of the treatment he would get and the possibility for promotion, so the Welfare Section was said to be the unit where Africans could advance the farthest (with advancement judged by Africans to be on the one single and critical criterion, salary).

Welfare Section personnel worked in one of two divisions, Case Work or Group Work. Case workers dealt more closely with specific individual and family problems than did group workers and thus were more frequently in touch with alcohol-induced or alcohol-aggravated problems. Group workers consciously sought to provide alternative social activities for township residents, particularly for youth of about the age when the mysteries and realities of beer-garden life were beginning to exert a strong attraction.

Point of View: A Case Worker

An African case worker reviewed for me the advantages and disadvantages he saw in the beer industry and then enumerated the types of alcohol-related problems with which he customarily dealt. Among the advantages, he noted the fact that Africans coming to urban areas seemed not to be able to occupy themselves gainfully, in contrast with the way Europeans organized their time, and thus the beer halls provided a place to spend leisure hours. He suggested other advantages that have been noted, including the way people could meet together to keep alive their own traditions of dancing, singing, playing musical instruments and dressing; the social functions of circulating news and information, finding a wife or learning from one's peers about how to obtain a house; economic functions such as learning about jobs or how to obtain health and welfare benefits; and the fact that the price of beer had remained reasonable.

On the negative side, he pointed to the problems of excessive drinking and how the easy availability of beer contributed to it:

> "If it wasn't so easy to buy beer, the inducement wouldn't be there. Practically speaking, there is literally beer everywhere and it is easy to get. Any time of the day, if one beer garden or outlet is closed, another is open, until about 10:30 PM. People coming from the rural areas have many things to choose from. Unfortunately, many of them have chosen to drink all the beer possible. Even people who have no desire to drink find it a temptation and attraction. Today many people drink: young people, even church people who didn't used to drink a few years ago.
>
> "I'm of the opinion that because the Africans can't use their leisure, and they don't think of taking a vacation leave, the tendency is to find all their leisure in the beer garden and to use all their money for drinking."

The case worker pointed out the dissonance created for men by the fact that an African man is expected to keep out of the house, and that about the only places where an urban man could spend his time in the company of other men were at sports events or in beer gardens. At the same time, the feeling persisted (as evident in the comments of the missionary quoted previously) that through his drinking the urban African was playing right into the White man's hands.

In reviewing the problems he faced in his daily rounds, the welfare worker expressed his concern for a general lowering of morals and an increase in the amount of crime and fighting. In the beer-garden climate, social relationships were freer. "People can do a lot of evils and pretend that they do them because they are drunk," he said, but he emphasized that beer drinking was not the only cause of problems. "If you took the beer away, you would still have these problems, but not to such a great extent." He cited other major factors contributing to neighborhood friction, including a general problem of intertribal jealousies and specific disputes arising over the behavior of families, children and dogs. Witchcraft provided another source of problems for case workers, he explained.[1] In-

[1]See also Gargett (35, *pp. 121–122*) in this regard.

ternal family problems were occasionally attributed to the evil intent of one's neighbors. "Sometimes the witch doctors stir up this kind of trouble 'for money'," he said, but suspicion could also be aroused by a seemingly innocuous act like throwing washwater into a neighbor's yard.

The case worker systematically described the types of drinking-related problems with which he dealt, most of them involving problems of family disorganization:

1. *Nonsupport.* "Families with insufficient food in the house, the wife and children insufficiently clothed, the rent or medical fees unpaid, or the children have failed to attend school because there is no money for school fees."

2. *Assault.* "The wife comes in to complain that she has been assaulted. Her husband came home drunk, she reports, and when she asked him where his pay was, he hit her." [To illustrate this point, the welfare worker produced a metal axe and a collection of wooden knobkerries gathered subsequent to such assaults. Although it was illegal to carry such weapons, they could still be kept in the house for self-defense.]

3. *Desertion.* "A man may desert his family in one township and go to another township to live with a woman he has met."

4. *Adultery.* "In one case, a man who was drunk brought a girlfriend home from the beer hall. He introduced the wife to her, forced the wife to accept the girlfriend, and then forced them to sleep in the same bed with him."

5. *Parties in the homes.* "Sometimes people buy beer and bring it home after the beer halls close. They have a party, turn the radio on, and keep the kids up. Then the neighbors complain. We are told that the kids who come from homes where there are beer parties do not perform well in school."

6. *Shebeens.* "Some people buy European beer and liquor and serve it at home. Young girls—the school leavers—serve the beer or dance with the customers. These young girls, starting at 16 or even 15, also begin to drink. Often they get pregnant and they don't even know the men who got them pregnant."

7. *Health.* "Sometimes it's the husband who calls in the Welfare Officer because of having to live in a dirty home. Then we call in the Health Officer. Here the culprit is the wife—she is in the beer hall as soon as it opens each morning, and she doesn't look after the home. And then the husband may start suspecting the wife, because he wonders where she gets the money she spends on beer."

8. *Alcoholism.* "There are people who become hoboes. They have been left by their wives, they cannot rent a house, and so on. Yet just how they are a problem is not clear to us. A doctor at the African Hospital may send me a note, 'This man is suffering from pellagra—can you get him a job?' or 'Can you assist this person in obtaining food?' There is a lot of malnutri-

tion. I think this is going to become more of a problem, but it is not one yet. There has not been much call for our services in such cases. The extended family and the kinship system of African society cover it; other people look after an individual who needs help."

9. *Jobless people.* "When they start drinking, people who are out of work don't go out and look for a job. As the rate of drinking goes up, this may become a more serious problem."

Point of View: Group Workers

Along with limited but valiant attempts to help some urban Africans occupy their time gainfully, ranging from art-and-craft centers which kept a few people employed to programs which authorized women vendors identified as welfare cases to prepare and sell food in beer gardens, the welfare division also offered a variety of club activities especially designed for urban African youths. These activities were managed by group workers assigned to the Youth Work Section. As the youth workers perceived their role, beer gardens represented "the enemy," and youth programs were intended to offer viable and attractive alternatives to the beer gardens, particularly for youths under, or just at, the legal drinking age. These youth leaders, themselves Africans, were professionally concerned with drinking among the younger generation, as comments volunteered in interviews with several male youth workers suggest:

Youth Leader One: "Drinking so young—even in the ages of 12–15—is a worry to the African society. A child that age should get an education. He can't do that if he has beer in his head. I don't know why they drink when they are so young. We say that drinking makes a man feel young. But why stimulate a boy of 12 years old? He already feels good. And when you are young, you are still soft everywhere. Drinking is something for older people!"

Youth Leader Two: "I don't drink and I don't spend much time with people who do, so I don't know much about drinkers. Some get aggressive, but that's a minority. Most get gay. That's what most people drink for. In a pub there may be more dancing, or else people just tend to be conversational.

"Our Youth Centres provide some weekend activities for the 16–23 year olds. On Sunday afternoon they have ballroom dancing. We have no problem with their drinking. When they drink, they don't come around here. When they are here, they're jolly good fellows.

"One of the problems with the young men when they drink is their aggressiveness. They tend to want to prove what they are not, and to show more fighting strength than they have. They get into fights and treat everyone badly, especially the young ladies.

"The number of girls in the age range of 12–15 who drink is rising. No parent would knowingly allow girls that age to drink, but they do. The regulations are not enforced. You can see girls drinking right outside the beer halls on weekends. Their older friends bring beer out to them. Or the girls go right into the pubs. They get in with no problems. Parents don't seem to chase up where the girls go.

"I wouldn't say close the beer halls. But it would help if something were done to keep the youngsters out. Personally I'd say age 21 is reasonable; 19 is too young. I wish the regulations could be enforced. Today one is bound to find 30–40% of patrons in the pub under age. Except at the cocktail lounges, they never do anything to enforce the regulations."

Youth Leader Three: "These boys interested in drinking lose their respect. Most of these young fellows do drink. I haven't found the answer to why they drink. Where they get the money for drinking and smoking, I don't really know.

"It's hard for telling the ages of letting people into the beer halls. A boy of 19 looks the same as a boy of 22, and there is no way to prove age. It is a serious problem. An adult association we have formed is looking into this problem. They are trying to find a solution for getting rid of these boys in the beer halls and to enforce the original rule of 'no one under 19' in them.

"Parents themselves need to be interested and educated. Parents should be educated not to send children in to buy beer. Sometimes the parents even go in with no money and they see the youths there and have them buy beer for them."

8

Professional Points of View: The Police

RHODESIA'S national police force, the B.S.A.P., derives its name from the early days of the British South Africa Company which pioneered the development of European interests in southern Africa. Although the force was rooted in British tradition sparkling with polished leather, shiny brass and stiffly starched khakis, and still traveled unarmed in the normal course of duties, it was no longer either British or South African. Its effectiveness in Rhodesia in preserving the law of the land was acknowledged by Whites as well as Blacks.

Under the conditions of political self-consciousness prevailing in Rhodesia at the time, police power represented a double-edged sword. As professional policemen, the European and African members of the B.S.A.P. heralded law and order in a country that could not help but take pride in its low incidence of crime. As local emissaries of national policy, however, the police represented a more potent role than mere servants of the people.[1]

Since the larger municipalities in Rhodesia were deeply involved in the liquor business, they were in one sense ritual adversaries toward formal police obligations with law and order. Municipal officials were sensitive to police views and especially to police complaints regarding liquor sales in the townships. An official in the Housing and Amenities Department explained,

[1]White Rhodesians appeared sensitive about the issue of whether Rhodesia was in fact a police state. In proof, they typically pointed out that one seldom saw policemen about, somehow confusing the issue by equating the prevalence of policemen with the presence of a police state. With similar logic they might also have pointed to the exceptional efficiency, courtesy and morale that appeared to characterize the force. The real test, however, was not in the size or curbside manner of the police but in the extent of discretionary police power and the ultimate responsibility for the consequences of police action. One prominent White scholar and former resident was quoted from London as having stated that Rhodesia was a police state "for Africans and non-conforming Whites."

"We're trying very hard to build up a relationship with the police. But their general feeling is that we sell too much beer. They remind us that liquor is at least a contributing cause to problems of robberies, casualty assaults, and the like, and that there has been a steady increase in these things. And they're right. That is, the increase corresponds exactly with our graphs of improved sales. But they also appreciate that we sell a good product and that if we didn't make this beer, the people would get something worse to drink. They agree with us—but they can't help throwing in their point of view."

An Interview with the Police

Although the first off-hand comment made to me during one extended interview with three European members of the B.S.A.P. force serving the African townships in Bulawayo was, "I've never seen a place with as many beer gardens as they have around here," the police did not insist that the municipality sold too much beer. They did admit, "We don't altogether approve of all these outlets," and noted, "If you control the drinking, the crimes go down—anywhere in the world." Yet almost in the same breath one official acknowledged:

"From a common-sense point of view, the Africans would probably brew beer if they couldn't buy it. We can't just turn the clock back and reduce the facilities. But if we were living in Utopia, and everybody was a little more relaxed and happy, then maybe every time an African had two shillings in his pocket he wouldn't have to go right off to spend it in the beer garden. It's going to take donkey's years to get them to drink less."

In spite of their recognition that the African has always been a beer drinker, and that if beer is "overcontrolled" or not made available to him, he will always find ways to obtain it or brew it himself, their point of view as policemen left no doubt in their minds that alcohol was responsible for a lot of crime. Several specific alcohol-related police problems were enumerated:

1. Illegal sales outlets—specifically the shebeens—and the fights, prostitutes, loud music, and gambling associated with them. "For us, the problem is the lack of control over it."

2. Drunken persons as victims of thug attacks or robbery, particularly individuals too drunk to be able to defend themselves or to be able later to indentify their assailants.

3. Drinking-related personal accidents and deaths, including such commonly recurring problems as fractured skulls and somewhat unique acci-

dents like "death due to asphyxia" in which a drunken person staggers and falls into an open storm-water drainage ditch, vomits and then drowns in his own vomit, or two recent and similar accidents where men had returned intoxicated to 3d-floor rooms in the worker hostels and, while attempting to urinate out of upper story windows, had fallen to their deaths.

4. Fights and stabbings. "It's a known fact there's more assault here than in Salisbury, but we can't seem to trace any pattern unless there's just more intertribal rivalry here. It takes nothing to get these chaps fighting. Here are some reasons we have gathered as the cause of assault," explained one officer, reading from a police report: " 'nudging another man in a beerhall queue'; 'stepping on a man's foot in a beerhall'; 'refusing to give a man a light in a beerhall'; 'refusing to give a man a cigarette in a beerhall;' 'dancing to someone else's music in a shebeen.' "[2]

5. Drunken driving.

Of all their problems, the thug attacks by small gangs, usually of relatively young persons, seemed to be the most disconcerting to those charged with law enforcement. The fact that such attacks were so often directed against persons returning home alone after an evening of beer drinking had led police to draw their attention to the victims in what was otherwise a "needle in the haystack" problem of trying to locate assailants. One officer explained:

> "Our police patrols are mainly outside the beer halls. In an attempt to cut down this problem of thuggery, last year we actually posted uniformed men inside the beer halls. And we also do a lot of plainclothes work in them. We found that our patrols didn't make an awful lot of difference in the incidence of thuggery. We've kept it down, but we've not been able to reduce it."

Police in both Bulawayo and Salisbury were agreed on two facts and on the unspoken but, at least to them, obvious conclusion that one could draw from them: crimes of violence had been on the increase; drinking had been on the increase.

Whatever the problems in terms of absolute numbers, police officials noted that, by comparison with other urban populations, their problems of law enforcement were not out of hand. (They could not resist reminding their American visitor of crime rates in cities like Washington, D.C. or Chicago.) "The African is basically an honest person. If he wasn't, it would be an impossible task to police an area of this size. We have very few house robberies and thefts. Our

[2]Police explained that a patron wishing to have a record played paid 5¢ to the shebeen owner. The person who paid for the music could restrict for himself the privilege of dancing to it and might fight to defend that privilege.

crimes of violence aren't anywhere near alarming for a community of this size." The police quoted their own population estimate of 170,000 persons living in the Bulawayo townships, against which their cases of serious assault in 1969 numbered 571, an increase of 9 over the previous year; there were 223 cases of robbery by more than a single person, an increase of 7 cases from 1968. The number of deaths by assault in the townships during 1969 was 4.

The police had two primary targets toward whom they directed their efforts at crime prevention related to drinking: officials of the Housing and Amenities Department, and township residents themselves. Recent recommendations which they had made to officials included limiting the amount of draught beer sold for home consumption ("Any amount over 5 gallons is going into something more than a home party"), adding new illuminating lights at "dark spots" along frequented paths and roads home, and providing more facilities for the younger set, perhaps to include inexpensive cinema showings, more youth clubs and live music for contemporary dancing. The police felt that their advice to the Housing and Amenities Department was carefully considered, citing in support that an earlier recommendation regarding the need for a tougher policy on shebeen operations had resulted in a new housing regulation that called for a warning on a first offense and an eviction on the second (for purchasers and renters alike).

Public information efforts directed toward the urban African clientele took a characteristically European middle-class form—printed announcements of advice in a regular feature in the monthly magazine *Masiye Pambili* ("Progress," or, literally, "Let us go forward together") printed by the Housing and Amenities Department and distributed free within the townships to a guaranteed circulation of 21,000 copies. The advice printed in the column was relevant and helpful, but the police confessed doubts that the column was read by more than a few people and assumed it was heeded by still fewer. My attention was invited to a file containing copy prepared for the September 1970 issue of *Masiye Pambili*. In final publication the article was sandwiched between a churchman's column, "Umfundisi [the minister] Speaks" and a full-page insert prepared by the Bulawayo Family Planning Association. It read, in part:

> "I would like to repeat my advice to you regarding attacks by bad people. . . . You should avoid taking short-cuts along footpaths through bush areas. Keep to the lighted streets, and try to have a companion with you.

Thugs look out for solitary persons, because they can go about their cowardly work without being observed.

"Please do not drink so much beer that you are incapable of defending yourself, or identifying your assailants. The thugs prey on such people, and it makes their task so much easier.

"Since my last warning many people have been badly hurt and they are sorry that they did not listen to my words of wisdom."

The column was headed with and signed by the word "Umgcini," an Ndebele word describing a person to whom one can entrust property.

At the conclusion of one of the interviews with police, an official provided an interesting distinction between police awareness and police authority: "These days you can't ride roughshod over people, even if it's for their own good. If a man wants to drink himself to death, we have no control as long as he doesn't break some other law. Only when he steps out and commits an offense do we become interested."

9

Professional Points of View: Physicians

THE BULAWAYO TOWNSHIPS and surrounding tribal trust lands, totaling a population estimated for public health purposes at 1,300,000 Africans, were served by the large Mpilo Central Hospital located at the edge of the townships.[1] My initial query there was to get a medical opinion regarding the nutritional value of African beer. The European physicians with whom I spoke were interested in a host of other medical problems associated with, or suspected of being associated with, the drinking of African beer. The following notes have been extracted from extended conversations with several staff doctors, incidental comments by other doctors, and references in medical literature dealing specifically with problems related to the drinking of African beer.

Each doctor advised me to proceed with caution in trying to represent the medical point of view. In discussing medical aspects of African beer, they pointed out, it is difficult to distinguish factors related to alcohol content, factors related to nutritive content and factors related to external circumstances like beer-hall fights. They also warned that their opinions were often impressionistic, based on clinical observation but without the luxury of time or research assistance necessary to pursue systematic inquiry. The latter caution is hardly disconcerting to the social scientist, who assumes that medical "impressions" play a more important role in the urgency of coping with medical problems than doctors usually care or dare to admit. However, attention has been given to include medical opinion and also to identify available references to research studies.[2]

[1]Mpilo Hospital had almost 800 beds. The name comes from the Ndebele word for health. The complete medical facilities of the hospital were augmented by urban clinics and rural hospitals dotted throughout the region.

[2]I wish to acknowledge the special help of six physicians in the preparation of this section and critical review of early drafts of it: W. Fraser Ross, Dean of the Medical School, University of Rhodesia; W. M. Buchanan, also of the Medi-

The areas of greatest medical concern regarding urban African drinking were trauma (used in the medical sense to mean personal injury,) pellagra (a food deficiency disease), liver disease and iron overload. The extent of beer-related problems from the medical viewpoint was far-ranging. Topics are presented here in a sequence that doctors themselves used to organize their comments, beginning with consequences of drunken behavior and proceeding to nutritional problems and then to disorders of the neurological, cardiovascular and gastrointestinal systems. The doctors also noted some missing syndromes among their African drinkers and made some observations regarding Africans as patients. I have added a note concerning the role of African beer in European folk medicine.

Consequences of Drunken Behavior

"About half of a surgeon's job in this hospital is patching up people related to problems with alcohol," summarized one Mpilo Hospital surgeon. His colleagues appeared to be in agreement with the undue extent that trauma was a cause of hospital admissions in the African community, a cause which from both medical and humanitarian perspectives represented what one doctor described as "a potentially avoidable source of human suffering and waste of resources" (136, *p.* 522). Trauma had, in fact, accounted for one-fifth of all admissions to Mpilo Hospital over the course of several preceding years (108), a pattern which doctors felt was still in effect and in some ways had become more serious as new forms of personal violence (e.g., knife wounds) increased in frequency over older ones (e.g., being hit over the head with a knobkerrie).

Admissions for trauma were three and a half times greater in

---

cal School Faculty; and Norris M. Baker, Terry Buchan, I. Rachman, and J. E. P. Thomas, of the staff of Mpilo Central Hospital. Readers of early drafts pointed out that the relatively greater attention given to the medical section may inadvertently "give the impression of an alcohol-sodden population." That is not the intention here, although I do confess to having found the medical problems, and the sometimes divergent points of view about them, an interesting part of the research. Readers may also note that I have attended to the range of medical problems but have not attempted to report rates at which the problems occur or provided comparisons with other populations. Such data accumulate slowly, especially where rural indigenous populations are concerned.

number than the next highest category of causes. The nature and extent of the violence had created lasting impressions on members of the medical staff:

"My impression is that 75% of personal violence treated here is alcohol related."[3]

"Visiting doctors from England have told us they thought the thuggery was tough in Birmingham until they came here."

"The extent of head injury is colossal among these chaps. They're always bashing each other on the head. People are brought in who have been hit over the head by a gang of youths, or kicked in the stomach, or stabbed."

"On a fine summer night, two or three dozen may come through. You'd be impressed by the number of cases obviously linked with alcohol."

"You ought to spend a Friday evening at the hospital in Casualty. There's blood all over, and drunks lying everywhere. It's like a casualty clearing station on a battlefield. The last Friday of the month (payday) is the worst. Everybody dreads being on duty Friday and Saturday nights."[4]

In one attempt to track down the incidence of stabbing injuries of the chest and abdomen, a staff physician had observed the high correlation between the times of stabbings and the hours when beer gardens were open or the days when drinking was heaviest (136). The relationship he demonstrated was not convincing from a strictly analytical point of view. The author observed, for example, that although the beer gardens closed "at sunset or shortly thereafter" (*p. 524*), the admissions of stabbing injuries reached a peak in the period 10 PM to midnight and subsequently remained rather high until 6 AM. Similarly, the fact of the greater number of stabbings occurring

[3]A survey conducted in 1962–63 showed that alcohol had been consumed prior to injury by 60% of the admissions classified as assault (107, *p. 755*). Data were not available to show whether the proportion of alcohol-related injuries had risen in recent years or whether doctors had an inclination to overestimate the proportion of admissions who had been drinking. Possibly both factors were at work.

[4]I visited on a Sunday night on a payday weekend. "It's pretty quiet tonight," commented several of the staff members on duty, "but you should have been here *last* night." In addition to numerous persons treated and released, the previous evening had contributed five admissions to the wards: two skull fractures (one made by a knobkerrie, one resulting from being hit over the head with a stone), two stab wounds, and one person with a broken arm and leg resulting from a road accident.

on Saturdays, Sundays and Mondays could certainly be correlated with phenomena other than drinking hours. If such correlation studies do not "prove" anything, the fact remains that a staff doctor at the hospital had conducted the survey, his colleagues were conversant with its conclusions and, like similar surveys at the hospital (107), it tended to validate a general impression held by hospital staff that certain types of admissions were associated with urban drinking patterns. The author of the stabbing survey observed, for example, "If there is an associated head injury or respiratory embarrassment, the passage of a wide-bore stomach tube is essential and it is not unusual to remove pints of beer from the stomach of these patients with great benefit" (136, *p.* 525).

Doctors cited numerous ways in which drunkenness contributed to medical problems. Bone fractures, head injuries and stabbings, with a variety of instruments ranging from flick-knives and butcher knives to screwdrivers and sharpened bicycle spokes embedded in wooden handles, occurred along roads and footpaths and in dwellings as a consequence of quarrels over gambling and women (136). Road accidents in the townships were numerous, with road traffic on narrow streets complicated at night by myriad bicyclists, most of whom took their chances cycling after dark without lights or even tail reflectors. Automobile-bicycle accidents were an obvious source of trouble, but so also were cycle accidents involving pedestrians or other cyclists in what doctors described as the accident-prone behavior of persons who had been drinking. Lying comatose in a drunken state following a drinking bout had caused deaths by exposure and was felt to be the cause of lung infections in which a person's ordinary coughing and choking reflexes failed and he inhaled vomit or other infectious materials ordinarily excluded from access to the respiratory system. The African population was characterized as "very prone" to lung infection, including not only lung abscess but also pneumonia and tuberculosis.

Carbon monoxide poisoning resulting from having a fire burning inside a closed room was also cited as a consequence of drinking. Doctors felt that individuals returning home at night in a semi-drunken state were inclined to light fires for warmth and then lose consciousness before attending to ventilation, although other observers believed that such behavior was related to the failure of many Africans to realize that "ventilation does not occur in city dwellings the same way it does in grass huts."

Nutritional Problems

"The beer has plenty of calories. But I'm not so sure of its vitamin content," one doctor explained. Another suggested that an individual's pattern of drinking affected the extent to which nutritional deficiencies in the body of the beer might become nutritional deficiencies in the body of the drinker: "I think the beer must have a lot of nutritive value. After all, some of these people live on it for years. It is not a balanced diet, however, and if they drink only beer they run into trouble. But for the small number who drink beer every day of their lives, there are great numbers who 'go on a blind' only once a week or once a month, and for them nutrition is not a problem." An apparently changing climate of medical opinion concerning the nutritional value of African beer,[5] as suggested in the following excerpt from a medical journal, seemed to turn on this very point, whether man can live by beer alone (37, *p. 13*):

> "We used to think that African-brewed beer is rich in vitamins and very nourishing. For many years now, mine authorities have issued beer as a ration to their workers; we never realized that if taken in excess, such as upwards of one gallon a day, the beer might affect the health of the individual."

Physicians reported that malnutrition was not widespread in the African population although cases of it persisted among both chil-

[5]According to members of the research team at the Alcoholism Research Unit, the nutritional value of the municipal beers had declined in recent years, as cheaper ingredients like maize grits had been substituted for sorghum in South Africa and Rhodesia (102, *p. vii*), and they formally recommended that the beer be enriched, particularly to help rectify vitamin deficiency among the African population (*p. 36*). This appeared to be a delicate issue about which little information but much personal opinion was volunteered. Salisbury's Liquor Undertaking Department capitalized on the recommendation by introducing a new and "enriched" beer—at enriched prices that made even dedicated employees wince at the blatancy of the profit-making motive. Officials in Bulawayo claimed that their vigilance in maintaining the quality of ingredients in their beer was contributing to its increasing production costs. "We're using a wider variety of products and more expensive products than any other brewery—and our beer costs more to produce than anyone else's," I was told. References to research conducted in South Africa into the nutritional value of African Beer can be found in Novellie, "The food value and the intoxicating fractions of Kaffir Beer," in Nelson et al. (90).

dren and adults. "If they'd stick to their native diet of milk, maize and meat, they'd be all right," insisted one doctor, "but we sometimes find protein-undernourished babies because of the overdependency of these people on breast feeding." He also cited the practice of sending children out into the rural areas to be reared by other relatives, particularly grandparents, where they might be fed a diet limited to a mixture of maize meal and water. Among adults, malnutrition was reported among two types of cases, psychotics who refused to eat, and alcoholics suffering an over-all dietary deficiency rather than a specifically diagnosed one. Most of the forms of malnutrition which doctors discussed were related to specific vitamin deficiencies.

Pellagra (literally, "rough skin," one of the distinctive signs in its diagnosis) seemed the most prevalent of all the nutritional problems which doctors linked to the effects of excessive beer drinking (Plate 12). Gelfand (37), commenting on alcoholism among Rhodesian Africans, reported enthusiastically, "Whereas I formerly held that pellagra—a very common disorder in the African—was the result of a diet too rich in maize, we are now finding that practically every sufferer from pellagra is an alcoholic." His colleagues expressed hesitation about appearing to lay the cause of pellagra so singularly on beer drinking. "Pellagra is always a problem in a maize-eating population," observed one doctor. Addressing himself more directly to the relationship between pellagra and the consumption of alcohol, Ross[6] noted, "While I think it is true that many alcoholics might well have pellagra, I do not think it is true for one moment that most pellagra sufferers are alcoholics."

Speaking specifically to the vitamin deficiency leading to pellagra, one physician posited, "African beer is supposed to have a lot of vitamin B, but it may not have vitamin B_1." My check into available but apparently outdated medical dictionaries revealed some equivocation about precisely which components of the vitamin B complex were deficient in pellagra but agreement that nicotinamide (vitamin B_3) was the principal factor. The identification of actual cases of pellagra appeared to be less subject to ambiguity than the precise determination of its cause, and doctors treating African patients routinely looked for the telltale signs of this endemic disease. Proper diagnosis did not necessarily assure a cure among heavy beer drink-

[6]Ross, W. F. [Personal communication, 1971.]

ers: "Once you see a pellagra case, you'll probably see him again and again, because he simply won't give up his alcohol."

Scurvy was another consequence of a vitamin-deficient diet diagnosed among beer drinkers, related in part to the observation that alcoholics tended to omit citrus fruit and green vegetables from their customary diet. Scurvy was also felt to be related to an accompanying phenomenon, iron overload. Excess iron can oxidize the ascorbic acid (vitamin C) present in the body, thus producing the effects of scurvy (e.g., bleeding from gums and other tissues, and osteoporosis —thinning of the bones, particularly of the spine—which some doctors had diagnosed among African beer drinkers) in spite of an apparent intake of ascorbic acid (7, 117). As its name suggests, iron overload is a problem of excess rather than of deficiency, and for analytical purposes is discussed separately below. From the clinical point of view, vitamin deficiency and iron overload were often diagnosed as complementary and complicating consequences in patients with histories of heavy drinking.

Neurological Disorders

Epilepsy seemed to come to mind first among the neurological disorders mentioned, not because doctors suspected that drinking causes epilepsy (quite obviously it does not, since children and nondrinking adults the world over are subject to the seizures, at an estimated incidence of 1 to 2% of any population), but because they felt there was some relationship between drinking and seizures among the epileptics in the African population who did drink. "Although I have no factual basis for arguing that the extent of epilepsy is higher in this population," said a doctor particularly interested in the subject, "I do know that it is one of the single most common diagnoses of our medical wards. We have been trying to ascertain what part alcohol plays in precipitating the fits. We know that once a person is prone to epilepsy, drinking can aggravate his attacks." Three alcohol-related situations were described in which drinking provoked epileptic attacks, all of which contributed to the impression doctors had that there was a high correlation between having been on a beer drink and having an attack. First, and probably most important, was a reduction in blood sugar (hypoglycemia) caused by an excess of alcohol and a failure to ingest other foods. This sudden reduction of blood sugar could provoke a fit in an epileptic, but even in nonepileptics it accounted for some admissions of patients brought to the hospital in a state of coma anywhere from 12 to 48 hours after a drinking binge. Second was fluid overload, a consequence of the sheer volume of beer consumed by many of the drinkers. Third were fits which occurred while a person was acutely intoxicated. Epileptics who had sustained a head injury prior to their first seizure, a type of injury frequently associated with drinking and drunkenness, accounted for 8 among 132 new cases of epilepsy recorded in 1 year at Mpilo Hospital. Alcohol was reported as a precipitating factor in at least some attacks among 6 patients who were reported as heavy drinkers, 2 of whom had fits only after drinking bouts (99).

In spite of the medical interest and the commonness of epilepsy in the African population, doctors recognized some peculiar difficulties in getting complete and accurate information about it. Among Africans, they explained, and especially among their less sophisticated rural clientele, epilepsy was often considered to be peculiar and unique to the African population, a problem completely outside the pale of Western medicine. In one medical study, 59% of the sufferers in a rural area near Bulawayo reported that their epilepsy resulted from their having been bewitched (77). Thus much of the behavior of epileptics before, during and after seizures was never entered into medical histories. Did drinkers have more frequent attacks? Did more epileptic seizures take place in connection with weekend drinking? These were questions over which the doctors puzzled.[7]

Alcohol intake plus nutritional deficiency was felt to contribute to general neurological disorder through a toxic degenerative process which could affect the brain ("chronic brain rot") and spinal cord. Over a period of time, doctors reported, alcohol "blunts the mind," and the "sort of 'spastic' behavior" characteristic of the comic inebriate became the ordinary behavior of a person in a relatively sober state as well. Memory impairment and mental deterioration (Korsakoff's psychosis, Wernicke's encephalopathy), temporary mental confusion (periodic African psychosis) and depression and persecution complexes were all felt to be related to drinking. Patients were oc-

[7]Medical and anthropological investigators interested in studying epilepsy among Bantu peoples will find a study conducted among the Baganda by John Orley helpful, including his Appendix C, "A guide for investigation of beliefs about epilepsy" (94).

casionally misdiagnosed as "mental cases" when symptoms such as the dementia associated with pellagra, the hand tremor and general confusion associated with liver failure, or alcohol-provoked seizures, were misinterpreted in relation to the whole disease complex.

Cardiac, Pulmonary and Gastrointestinal Disorders

"Alcohol in high quantity is a direct toxin to the heart," noted one doctor during an interview, "and I think a disproportionate number of our male heart-disease patients are heavy drinkers. There are quite a lot of African alcoholics with heart diseases who would be all right if they didn't drink so heavily." Another doctor estimated that some 20 to 30% of the heart disease treated at the hospital was for cardiomyopathy. "It's similar to what they call 'beer-drinker's heart' in many parts of the world," he said. "It's a disease of the heart muscle, and you often find it among alcoholics. Yet in some localities they can trace it to a parasite, and in others they link it up with nutrition. It's a bit of a mystery."

The medical observation of the susceptibility of the African population to lung infection generally, and to lung abscess associated with inhaling vomit, has already been mentioned. "We see quite a few such cases in our African population. In most of them, a chap has been on a drinking bout. It's a rarity to find this in the European hospitals."

Heavy drinkers anywhere are subject to certain common gastrointestinal problems. Doctors mentioned problems of bleeding in the esophagus, resulting from the tearing of tissues subsequent to heavy drinking and vomiting; gastritis subsequent to heavy drinking; and pancreatitis, a well-recognized problem among European alcoholics but a relatively rare one among Africans—"Although we're beginning to see a few cases here among the heavy drinkers."

The problem that most intrigued doctors in terms of African drinking was one that affects the telltale organ of heavy drinkers everywhere—the liver. The complicating factor in liver functioning for the African drinker was not simply the endless task of metabolizing large quantities of alcohol but that African beer was believed to contain a rather large quantity of absorbable iron. The alcohol in the beer further predisposed the system to high iron absorption. The result, at least for some heavy drinkers or for those whose systems showed a proclivity for iron absorption, was a medical problem known as siderosis or iron overload, a condition sufficiently common among members of the indigenous population of southern Africa, particularly among men, to have its own special name, Bantu siderosis. Because the liver is the body's single most important organ for storing iron, its iron content is frequently used as an indication of the body's iron storage level. The highest liver iron concentration in Africans had been found to be approximately 15 times the highest found in Europeans (8; 13, *p.* 5). The greatest total amount of iron found in the liver of one individual, a 55-year-old African man, was 32.9 g, more than 100 times the average amount in healthy Europeans (13).

Iron overload is not a problem restricted only to the liver. Excess iron deposited in the pancreas was believed to be one cause of diabetes; one doctor estimated that from 5 to 10% of the diabetic cases seen at the hospital were related to iron overload. Clinical impressions suggested that the heart and testes could also be affected. But it was in connection with the liver that iron apparently produced its most widespread consequences. "The pathologists tell us that you can actually *see* the iron deposits in the liver," commented one staff physician. "One study made in Johannesburg showed that there was several times more iron in the stool of African patients admitted on Monday than later in the week—sometimes into the hundreds of milligrams—probably representing the week-end drinking habits of urban Bantu."

Hard facts concerning iron overload were not readily available to hospital staff. The brewery did not know the iron content of its beer because no one had ever asked, although upon receiving a request for that information the brewers expressed confidence that their beer was "not excessively high" in iron. Earlier medical research had been carried out principally in South Africa. The hospital library contained at least one major reference to iron metabolism, including a chapter by Bothwell (6) reporting the South Africa research. Bothwell's article noted that while in the average diet a person ingests 20 mg of iron per day, a Bantu-beer drinker might consume between 50 and 100 mg of iron daily from beer alone. The iron intake among the Bantu was reported to be several times that of the White population. Furthermore, daily variations in the amount of iron intake reflected the prevalent Bantu pattern of heavy weekend drinking. Exercising some degree of caution, the report concluded, "Alcoholism may be an important aggravating factor in iron overload" (6, *p. 371*).

Six years and millions of gallons of beer later, doctors at Mpilo felt less hesitant about a positive relationship between the drinking of African beer, iron overload and a large proportion of the liver disease treated at the hospital. Although they were unable to quantify their observations, they felt that clinical observations provided ample evidence in this regard, and they were supported by a growing body of medical research into Bantu siderosis in Rhodesia (12, 13, 14). Nonetheless, they remained uncertain about either the long process of iron accumulation in the body or the contemporary sources of iron: "We used to think absorbable iron might come from the old iron cooking pots that they brewed the beer in, but these pots aren't used much anymore, and the problem is still with us." Most important, they had no way of distinguishing between cases of iron overload originating in rural areas, where brewing was still carried on in discarded 44-gallon petrol drums, from cases originating among urban drinkers whose beer was brewed in the stainless steel vats of the municipal brewery.[8] One doctor stated that it was his impression that most of the cases of porphyria he had seen were among patients from rural areas or who had resided in the city for only a few years and whose drinking had been done mostly in the country.

Iron overload was suspected by some physicians of contributing to two closely associated diseases common in the African population, cirrhosis of the liver and porphyria cutanea tarda.[9] Opinions varied as to whether it was also associated with a third, also a frequently appearing problem, cancer of the liver. "It's somewhat of a debate

[8]To everyone's surprise except the brewers', further inquiry revealed that brewery equipment included some iron tanks still in use among a majority of stainless steel ones. Preliminary tests conducted at my request showed a higher iron content in beer fermented in the iron tanks. Iron piping at the brewery and at the beer gardens was recognized as another probable contributor of iron, although in unknown quantity. Due to a lack of specialized equipment, I was unable to obtain a comprehensive analysis of Bulawayo's African beer. It was interesting to note professional commitments in the interpretation given to the laboratory analyses of the quantity of "available free iron," interested doctors finding "not negligible" the same amounts that brewery staff declared "negligible." Data available from an analysis of the iron concentration of 5 samples of Salisbury's African beer compared with 50 samples of home-brewed African beer revealed an average iron concentration almost 25 times greater in the home brews than in the municipal brew (13, *p.* 23).

[9]It should be noted that animal experiments had suggested that iron does not predispose to cirrhosis—Dunn (29) and Witzleben and Chaffey (137); but see also Orfei et al. (93).

whether cancer of the liver is related to iron or is due to some other cause. My own theory is that quite possibly it is iron excess," one doctor commented. A colleague rendered a different opinion: "The one authority whom I've heard on the subject said that cancer of the liver is not increased by iron overload; he felt there was no link between them."[10] In fact, as doctors cautioned time and again, the relationships between alcohol intake, iron overload and liver diseases are not well understood.

Drinking was believed to be a contributing factor in cirrhosis ("I believe we can relate cirrhosis to alcohol in most cases") and drinking African beer in excessive amounts presumably produced a double-barreled attack on the liver by introducing two substances potentially injurious if taken in large quantities—alcohol and iron. Patients with cirrhotic livers were susceptible to the further complications of porphyria and cancer. Yet among the patients diagnosed as having cirrhosis and cancer of the liver were young children as well as adults who stated that they were nondrinkers.

"I think that most of our porphyrics are chronic alcoholics and that their liver disease is associated with iron intake," summarized one doctor. The porphyrias comprise a group of diseases some of which are associated with liver disease. The one specifically associated with Africans who had siderosis was porphyria cutanea tarda symptomatica (28). This disease includes some rather discomforting symptoms: a darkening of the skin in light-exposed areas, especially on the hands and face; a cirrhotic liver; and, during acute episodes, the passing of a "port-wine colored" urine. "The Africans themselves know that the dark-faced ones are heavy drinkers. By the time people have 'gone dark' and finally come to see us, they usually have about a year left, although we do have one chap who claims he's laid off the beer and his condition has at least remained static for the past 3 years."

At home or hospital, discussions with staff doctors found them stopping to check their impressions with textbooks and recent jour-

[10]Our discussions almost invariably provoked subsequent searches of the medical literature. On the question of cancer of the liver, one doctor later reported, "Cancer of the liver commonly occurs in cirrhotic livers whether siderosis—iron overload—is associated or not." Bothwell and Finch (8, *p. 374*) stated that a link between siderosis and cancer had not been proved, although it is known to be more common in inherited iron overload or "hemochromatosis." Cf. Sherlock (118, *p. 371*).

nals or pausing to suggest other sources where further information might be available. During one extended evening discussion with three staff doctors, the host doctor interrupted the discussion to ask if he could offer his guests another beer. "By the way, I just looked up the iron content of European beer," he announced with a seeming note of relief, "and there's no iron in it at all!"

Missing Problems among African Drinkers

Along with the problems which they included in reviewing the medical aspects of African drinking, my informants also mentioned certain problems which were conspicuous by their absence in the pathology of drinking.[11] Beriberi, for example, a consequence of a thiamine (vitamin B_1) deficiency, was not among their diagnoses. Alcoholic peripheral neuropathy, a nervous disorder thought to be due to thiamine deficiency common in heavy drinkers, and a well-known symptom of alcoholism in the European population, was also uncommon. "I'm not sure I've ever seen a peripheral neuritis case which could be attributed to alcohol," one physician observed.

Most conspicuous by its absence was delirium tremens, the "D.T.s" of alcohol withdrawal. "It's very, very uncommon," suggested one doctor. "As a matter of fact, I don't think any of us has diagnosed a case of it here." Another doctor reported one positive diagnosis of "D.T.s" in 3 years. The apparent absence of this symptom of addiction so well-known to societies in Western civilization caused one author interested in alcoholism in contemporary African society to caution, "This is not to mean that chronic alcoholism is rare. . . . The pattern or picture of chronic alcoholism tends to assume a different one to that of the European" (37). On the issue of whether or not alcoholism followed different patterns among Europeans and Africans, doctors were not in agreement. One opinion was that "Alcoholics seem to be the same the world over. There isn't anything peculiarly African about them. Becoming an alcoholic seems to be completely random; you can't tell who will turn up. But at least African alcoholics admit that they drink. Even the women will admit they drink—they don't 'hide the gin bottle' the way European alcoholics do."

A psychiatrist who treated both African and European alcoholics

[11]See also Gelfand (37).

felt that there were important distinctions in the usual social circumstances between the two. He noted a high incidence of admissions of male European alcoholics, compared to a low rate of alcoholic admissions of Africans (although with only two psychiatrists seeing African patients in Rhodesia, facilities were hardly comparable). Among European alcoholics, he said, one typically found a history of heavy drinking dating back to adolescence: "They learned to solve their problems with alcohol early in life." A common pattern showed a personal history of job instability, marital problems, minor conflicts with the law and recurrent hospital admissions. At the present time, his older male European patients usually dated their heavy drinking back to the war, after which they "coped with and compensated for" continued drinking until they could retire with an early pension. They were spirits drinkers, possibly with a pattern of drinking in recurrent debauches. "Now he has drunk himself into a cirrhosis—he's an alcoholic." He added, "A lot of alcoholism is concealed here because heavy social drinking is so much a part of the European's life in Africa."

Africans admitted for psychiatric treatment, by comparison, usually had a stable work history and a fair income. They were beer drinkers only. The recurring factor among them was an element of social isolation—something happened to disrupt home or family life and they began going to the beer garden instead of eating. "But alcohol complicates their lives, it doesn't make them better," the psychiatrist pointed out. "I've joked with administrators in the Housing and Amenities Department that they ought to be glad any time that beer sales go down, because it suggests that the urban African is becoming better adjusted and less dependent on alcohol."

Some Observations of Africans as Patients

Doctors felt that their African clients had a high regard for modern medical practices—so high, in fact, that officials at Mpilo Hospital had levied a charge on patients in an attempt not only to produce revenue but to reduce the extent of visits deemed to be unnecessary. A new policy of charging a flat fee (outpatient calls, 25¢ per mother or child, 50¢ per man; inpatients, \$1, \$2 and \$4 for a child, woman, and man, respectively, regardless of length of stay; emergency, no prior fee required) satisfied the staff that it had accomplished its intended purpose when the staggering number of outpatients seen each day (an average of about 1000) diminished somewhat without a corresponding decrease of admissions among inpatients. Once admitted, Africans were described as being initially cooperative but exhibiting a tendency to become impatient under prolonged treatment either in the hospital or at home. One consequence of this impatience was that convalescing adults occasionally included a jaunt to the nearest beer garden in their own daily rounds:

> "A sick African doesn't like to stay in bed. He'd rather go sit in the sun. And once he gets out in the sun he's just as likely to bugger off to the beer hall. We have patients who will disappear in the middle of the day and come back later in the afternoon drunk as coots. We ask them how they get beer, since they leave the hospital without any money. 'People see that I'm not employed,' the African will tell us. 'Nobody will give me food. But they'll give me a beer'."

Regarding the medical aspects of other socially related problems, doctors stated that if there was any drug problem in the townships it was not widely prevalent and was limited to cannabis. ("I've never seen a case of it in a patient here. The police seem to know more about it than I do.") The venereal disease rate had been high in times past and had been sharply reduced, but was believed to be on the upswing again. One of the unique problems mentioned was the occasional need to remedy the remedies of native curers. Doctors were aware of the persistent belief, even among their younger, more urbanized patients, that sorcery and magic were still effective causes of illness. Staff physicians also had to recognize that simple illnesses were sometimes complicated by initial treatments prescribed by their indigenous counterparts. For example:

> "One of the medicines the witch doctors use in treating tummy aches is crushed wings and bodies of certain beetles. These contain cantharidin, which is purportedly an aphrodisiac and which for some individuals becomes a serious irritant to the urinary tract that can result in bleeding, severe toxicity, and sometimes death."[12]

Referring to their dealings with Africans with drinking problems, doctors remarked that they were sometimes amazed at how many of their patients reported curtailing or stopping their drinking when advised to do so. Consistent with their observation that African patients generally sought treatment rather than counsel, doctors noted that while they could never seem to convince patients to stop

[12] Baker, N. M. [Personal communication.]

drinking for the reasons that might appeal to a middle-class European (e.g., job security, effect of his drinking on other members of his family), their African patients, particularly the men, did express concern for their health and were often willing to give up their beer if they were told it was harmful for them. "I have the impression that it's easier to get an African to stop than to get a European to stop," summarized one doctor. "Maybe it's easier to put a person off liquor who drinks only on weekends rather than a person who drinks secretively and every day."

African Beer in European Folk Medicine

Almost anywhere except in my talks with churchmen or physicians, a European who heard of my interest in African beer was likely to inquire whether I was aware of the remarkable healing properties of the brew. The customary accolades included its general healthful qualities, its body, its wealth of vitamins, its quality as a "complete food." More specifically, it was recommended as "A great thing for blackwater fever ('double malaria') and for ulcers."

African beer was also appreciated as a general tonic, particularly good for "upset tummy and diarrhea." Houseboys from some European homes were reportedly sent to nearby beer gardens with surprising frequency to obtain beer in the interest of the health of local householders.

Doctors and Drinkers: A Footnote

In the course of investigating aspects of beer drinking from a medical perspective, I talked with scores of doctors, had an opportunity to accompany a few of them in consultation with patients, and read a sample of the vast literature which doctors read and have written regarding alcoholism and the treatment of alcoholics.[13] In that literature doctors *(1)* constantly tout the benefits of a multidisciplinary approach to alcoholism; *(2)* alternatively discover and reject new drugs to assist in the rehabilitation of their patients; *(3)* argue endlessly whether alcohol is a poison, a sedative, an anesthetic, a food, a fuel, or a combination of these; and *(4)* confess that except in clinics

[13]Anthropologist James Spradley has noted how recently alcoholism has come to have a medical identity as a disease but that we still **treat it** as a peculiar disease form that one *becomes* rather than *has* (126, **p. 66**).

specially designed to treat alcoholics, they'd rather not. My conclusion after extended conversations with doctors concerning alcoholism in African or European patients is that they do not find alcoholics professionally intriguing or satisfying, probably because they give a doctor little likelihood of effecting a successful cure.

This observation is not intended as an indictment. Indeed, European doctors working with African patients under the institutional auspices of a hospital the size of Mpilo may have been little affected by any professional disdain in spite of the large number of patients they saw and the lack of reciprocal interpersonal commitments between doctor and patient.[14] Doctors in private practice in the European sector of the community, apparently like their colleagues elsewhere in the world, were more concerned about, and perhaps threatened by, alcoholics among their patients: "There's little we can do for them, and most of them don't really want to be helped anyway. They take hours and hours of your time, and they never even pay you."

As already noted, doctors felt that African patients were far more likely to stop drinking when warned of potentially harmful results than were European patients. Nevertheless, I sometimes sensed a moralistic rather than a medical motive in the extent to which staff physicians spoke of advising their male patients to stop drinking and a tendency on their part to assume, unless there was compelling evidence to the contrary, that heavy drinking played at least a contributory part in many African patients' problems. Such an assumption had two potentially negative concomitants. First, doctors systematically ignored the possibility that there might be important differences between drinkers of municipally brewed beer and drinkers of beer brewed in rural areas. Second, by tending to associate certain types of admissions to the hospital with prior consumption of alcohol, and to assume that patients were probably heavy drinkers, doctors may have inadvertently but categorically assigned a majority of their patients the status of professionally unattractive cases. Regardless of personal as well as professional circumstances, it seems, *nobody* likes "a drunk."

[14]See Joseph (64).

10

Professional Points of View: Administrators and Managers

AT EACH BEER GARDEN, draught beer was delivered under pressure by compressors on the delivery trucks which pumped the bulk liquid up into the servery's overhead storage tanks. It was my impression that just as the beer was delivered under pressure, those who held responsible jobs directly related to the Liquor Undertaking also felt considerable pressure in relation to their positions. I include in this group everyone in administrative capacity, from the totally African coterie of bar managers to the totally White staff of senior administrators at Head Office. Like administrators in comparable positions but other settings (30, 139, 141), their success in their professional roles depended on their ability to mediate among a host of conflicting forces so as to avoid the kind of confrontation known locally as "the crunch."

Bar Managers

The African beer-garden and bar managers represented a status of "middle" managers whose complaints sounded remarkably like the collective complaints of such personnel universally: too long hours at too little pay, too much responsibility with too little authority, too little consideration for the special talents or requirements of their jobs (e.g., "I use my car so people will not attack me at night, but when I ask for allowance, they refuse"). Their work required them to be effective in multiple role playing. They had to be able to operate effectively within the framework of a competitive, materialistic, achievement-oriented society in fulfilling their official administrative obligations. "What kind of questions did they ask you in your interview for a promotion?" a junior grade manager inquired by telephone of a colleague at another beer garden during one of my interviews. At the same time they had to satisfy their African patrons that they had not become so co-opted by their administrative positions that they no longer felt themselves to be at one with their kinsmen and neighbors.

Managers of beer gardens and of bar-lounges alike were known as bar managers. Distinctions among managers reflected the hierarchical structure of the bureaucracy that had spawned them. Tiny differentials in pay validated formal status differences like one's grade (managers were of Civil Service Grades 6, 7 and 8), one's longevity as a manager, and the type of outlet and volume of business handled. Senior men (bar-lounge managers and managers of large beer gardens) constituted a group of reliable old-timers, but among the ranks of newly appointed managers and assistants turnover was frequent. This was due in part to the "get tough" policy established in the Liquor Undertaking section: whenever things began to look awry at a particular beer garden, such as problems with adulterated beer, unexplained drop-offs in sales, or unexplained discrepancies between beer received and beer sold, everyone on the staff, from the full-time bar manager to a part-time mug collector, was subject to immediate dismissal. During my first 3 months of fieldwork some 2 dozen African employees were dismissed, including 3 entire beer-garden staffs. Records during the previous 7 years showed an annual average of 29 employee dismissals and a range of from 15 to 57 per year. Such purges left everyone a bit shaken, Whites and Blacks alike. They epitomized the extent of the determination that officials felt toward protecting municipal beer (and profits), and the feeling of powerlessness on the part of any employee who might staunchly insist on his individual innocence.

I interviewed several bar managers and had the opportunity to speak briefly with many others. I explained that I was interested in their perspective and in the problems they faced in their work. The bar manager quoted below was a handsome Shona in his mid-30s; he spoke excellent English and, like many managers, said that he was fluent in both Shona and Ndebele. Bar managers and cashiers generally presented the image of bright, clean-cut and capable men. Their efforts in obtaining their positions and the salaries that accompanied them—ranging from R$50 to R$110 a month ($70 to $154 U.S.) were a function of ability and, among younger men, of formal educational attainment. Yet the salaries seemed generally unsatisfactory to them all, not because of the starting wages but because once a man reached the highest "notch" in his grade, his salary remained static. At the same time managers were, by the very nature of their work, reminded constantly of the amount of money they were returning to the city treasury. "It's financially adequate," explained one manager, "but according to the amount of responsibility we have, the pay should be better. We're looking after a lot of money. You might say, however, that there isn't any better pay than this for an African in Rhodesia."[1]

The excerpts that follow were taken from an interview conducted in the bar manager's small office adjacent to the servery.

"I held all sorts of jobs before I applied as a bar manager. When I first came to Bulawayo I took jobs as a houseboy, as a clerk, as a factory worker, and as a telephonist. I started as an assistant bar manager. The pay isn't good at all—I have a wife and family and other dependents to support—but only Welfare Section pays any better. This is a terrible job. We're accustomed to it, I guess, but our problems seem quite endless. For example, the African beer is kept in tanks calibrated with water, not with beer.[2] You never get the amount of beer you've signed for. You always get a deficit or a surplus. Sometimes the gauges get blocked. At the end of the day you may discover that you are short 30–35 gallons. If you report that you are short, your employers are very suspicious and they want to know why. Otherwise it would be a bad record. And your officials will never believe you, even if you tell them what is happening. This problem of deficits and surpluses goes on and on.

"You have to try to please everybody, customers and staff, and that's pretty hard to do. I have to keep the staff happy, and I have to keep them sober. We always have to be conscious of public relations. Sometimes we make charts or signs to remind the staff how to behave. I tell them, 'You just be cool—and remember the influence of liquor on our customers.'

"I think it's better to mix Shona and Ndebele on the staff. The City Council didn't design these beer gardens just for Shona employees. Shona are good workers, all right, but some of my staff are Ndebele. Anyway, we don't employ our staff—they send out the people we work with.

"Whether to drink with the customers is a problem. Sometimes you have to drink a little bit to please them, but usually you tell them you

[1]Although the African wage scale was depressed, there were better paying positions than that of bar manager. African teachers at government schools, for example, at least until a new wage scale was introduced in mid-1971, were paid on the same salary schedule as Whites and thus could earn two or three times that amount.

[2]"You just can't get a Kaffir to understand that a gallon is a gallon, no matter what you use to measure it," explained one White functionary responsible for monitoring the distribution and sale of the draught beer, but his tortuous explanation of the problems of accurately measuring active beer left me wondering whether indeed a gallon was always a gallon. Monthly records consistently showed an increase in gallonage due to fermentation.

can't drink when you are on duty. I don't drink. My mother used to brew beer, and it always made me sick. I never could drink it, even though I could tell whether it was good beer. I'm also a member of a church that doesn't allow drinking.

"Personally, I think the habit of drinking is going up too high. Some wives spend much time in the beer halls each day. They are regular customers. Some men drink 2 or 3 gallons a day—that's about a thousand gallons a year. On one recent Saturday 75,000 gallons of African beer were sold in the beer gardens.

"Bar managers don't go around giving people advice about drinking. That would be very wrong. A manager's job is to promote beer, to have people drink more, not less. If we started telling people they drink too much, either they would interfere with us or go to another beer garden. Actually we encourage them to drink. Even if the beer is bad, we say, 'Come on, man, it's quite good.' We are salesmen. When the beer is too thick, you're supposed to take it up and put it back in the tank. But it makes the customers unhappy when they see that.

"It's our duty to suggest certain things that would improve sales—like the special little area we have developed here for serving European beer. And we keep the bar clean and attractive. We keep the floors waxed and bring some flowers to make it look nice inside the servery.

"Customers can give you a bad time. Sometimes they start an argument, and they'll tell you, 'I already gave your cashier a dollar—where is my beer?' Or, 'Where is my change?' And people who know you from your neighborhood in the townships can come and make trouble for you in the beer garden where you are a manager. That would never happen to you if you were a clerk in an office. That's a time when you really need the support of your regular customers to defend you and say that you are a good manager."

Bar managers lacked consensus about their expected duties regarding the promotion of municipal sales. Unlike the man quoted above, another manager stated: "I'm not a salesman. The beer sells itself. Of course, we don't tell people to drink less. The City Council likes people to drink. But we don't look after the customers, they look after themselves. All we do is make sure they get the service the City Council wants them to have."

Opinions also varied among managers as to whether the extent of contemporary drinking exceeded the best interests of urban Africans. Several expressed concern about drinking among the young. Managers of bar-lounges had special help in this regard because they could post municipal police to monitor entrances. As one lounge manager explained, "We don't allow young people here. The young boys cause trouble. They get to fighting when they drink." But enforcing the age minimum in the beer gardens posed other problems.

Most gardens had multiple entrances or could not be closed at all, and many had well-worn routes for foot traffic crossing through them. A large number had "dry" canteens to which children came to purchase food items for their homes or bought candy and soft drinks for themselves. The manager of a beer garden located in an area with a large population of wives and children commented, "We have problems with young people. We cannot stop them from coming in. We have a dry canteen here, and they come in to buy sweets. Or they send in an old man to buy beer for them. It's serious, and if my officials find them, they will be unhappy. They also have a system of carrying beer out on the lawn. Outside the beer garden I'm not so worried—that's a police case."

"It's very open," another manager explained. "The young people come in and there's nothing to stop them. It's leading our youth to drink more and study less."

The collective comments of the managers suggested that they saw themselves as the "fall guys" in the Liquor Undertaking, sometimes being suspected of misdeeds and at other times expected to handle problems by tacitly recommended procedures equally suspect in nature. They also complained of the kind of inequities that seem inevitable in an organization the size of the vast Housing and Amenities Department:

1. "Reports are always supposed to 'jibe.' So if some beer goes over the top [freshly delivered, active beer often increased in volume after delivery, particularly in warm weather, and thus could occasionally overflow a tank], we have to make it up through the Buck meter" [i.e., dispensing beer so rapidly that the units do not completely fill, thus delivering a short measure].

2. "Sometimes you have to juggle beer around [opening the valves to pipes connecting tanks containing "old" and "fresh" beer and drawing on them simultaneously]. That's not legal, but we have to do it so we don't have to report that we had to throw away beer. We try to protect ourselves by pushing the old beer. We always use the old first."

3. "We try to keep the beer tasting good. One way to do it is to leave the tanks open at the top so the beer can breathe and not get sour. But official policy says that tanks must be closed and locked."[3]

[3]Brewery memorandum: "The main type of infection affecting African Beer is Acetic Acid Bacteria. For this type of bacteria to develop it requires free oxygen. Serious souring will take place if the beer is exposed to the air" (19, *p.* 5). Brewery officials and bar managers obviously held different views about the "breathing" requirements of beer.

4. "We work very long hours. We have to be at work at 7 AM to supervise the cleaning and to be on hand when the Area Controller comes to pick up the money. After our money is deposited, we have to wait for our delivery of beer before we can go home, but we don't get paid for that time while we wait." [Bar managers were considered to be off duty after morning clean-up until their bars or beer gardens opened. Managers at morning-opening beer gardens usually remained on the premises, but those who opened in the afternoon were off during the midday after they received their delivery. One manager of a late-opening site complained, "I've had to wait as late as 2 o'clock in the afternoon before I could go home."]

5. "We are the bar managers, but when we voice our concern, not much attention is paid. We have a very strong union, but in some cases they don't even listen to the union. Whenever there is a mistake in the beer garden, the officials have no confidence in us."

6. "I don't think there is any adulterating of beer going on at all. You can see for yourself the way the beer gardens are constructed that there are no means for bringing beer down. The City Council has tried all sorts of precautions to keep people from fiddling around. There's no chance for fiddling, but there's no belief and no trust in us."

Whatever their complaints and their discontents, not even the more anxious or pessimistic among bar managers were completely disparaging of their positions or totally overwhelmed by the problems they confronted. There were no complaints about the quality of the beer ("If the brewery keeps supplying us with good beer, we can sell it"), no complaints of gross inequities shown by senior administrators, and no outspoken complaints about the behavior of beer-garden patrons collectively. No bar manager stated an alternative job he would rather have had instead of his present one. Those more socially inclined (not one of the stated prerequisites for the position but clearly an asset for anyone who enjoyed the constant interaction in the beer gardens) found the job satisfying: "It's a good job. It teaches one to know how to look after people. You get respect, and you are known widely."

Area Controllers

The ever-increasing number of personnel assigned full-time responsibilities on the "liquor side" in the administrative units of the various township superintendencies and at Head Office was indicative of the ever-increasing prominence of the municipality's beer industry compared with other major facets within the Housing and

Amenities Department. Liquor administration had always been a focal role in the Department, but not until 1968 had the positions of "Adminstrative Officer, Liquor," (the "A.O.L.") and "Administrative Assistant, Liquor Undertaking," been established. For years the Township Superintendents of what had been Bulawayo's Department of African Administration supervised the beer gardens as part of their regular administrative duties. Now they had full-time administrative assistants known as Area Controllers who supervised staffs of African clerks and who were the officials (usually European)[4] directly and immediately responsible for all liquor operations —beer gardens, bar-lounges and bottle stores—within the township or a comparable administrative boundary. The 13 townships had been divided into 6 liquor-selling areas which coincided with major geographic administrative divisions.

A vast amount of beer was sold, and vast sums of money were collected, counted and deposited under the watchful eyes of the Area Controllers. The extent of the formal procedures to be followed and the control over the operation within each selling area were, to use a term suggested by one Area Controller, "staggering." Reports were prepared showing the day's performance at each outlet for each separate line carried. Accompanied by summary totals for the entire selling area, these were forwarded to Head Office in a constant stream of reports. Printed forms submitted daily included *(1)* Deposit Slip, City Treasurer's Department; *(2)* Bar Manager's Daily Report: African Beer Statement (Bottled and Packaged); *(3)* Bar Manager's Daily Report: Spirits Sales and Inventory; *(4)* Bar Manager's Daily Report: Beer and Stout; *(5)* Bar Manager's Daily Report: Bar Stock Return; *(6)* Area Controller's Summary Report: Daily Statement of A.B. (African beer) Sales; *(7)* Area Controller's Report: Daily Cash Summary; *(8)* Area Controller's Report: Daily Report on African Beer; *(9)* Area Controller's Report: Packaged African Beer—Daily Area Summary.

[4]During the period of fieldwork one African was assigned as an Area Controller. One Township Superintendent was an African. Africans of administrative grade were conspicuously absent among personnel assigned to top roles at Head Office. It should be noted, however, that African employees of the Department felt that they were well treated and that Bulawayo's Housing and Amenities Department was one of the few local or national departments in Rhodesia in which Africans had, since 1959, been accepted into professional grades in municipal service.

Area Controllers served as trouble shooters, general supervisors and inspectors. Their duties specifically entailed conducting major and minor spot checks (recorded on appropriate forms) which provided an immediate inventory of bulk beer, the item most subject to variation. Variation was recognized and accepted as an inherent problem in measuring active beer; variation itself was not necessarily looked upon with dismay. As noted earlier, the volume of fresh beer was likely to increase due to continuing fermentation after delivery, and thus a bar manager often had more beer to sell than the amount delivered to him from the brewery. The rapid turning of the Buck meter during a busy time resulted in a "short" measure which also contributed to surpluses. Thus a busy beer garden was expected to show a surplus. An Area Controller explained, "We allow up to 10% surplus. If there is too much of a deficit, we start getting suspicious. We expect the beer to show a surplus if things are busy."

Area Controllers were the watchdogs of sales as well as personnel. Taking a cue from the fondness for graphic presentation among personnel at Head Office to demonstrate ever-increasing sales, they dutifully plotted the sales record of each of their major lines and even recorded daily weather to help account for unexpected changes in the volume sold. Unusual performance at one outlet in contrast to trends at others precipitated suspicion, speculation and surveillance. Area Controllers reveled in periods of increased sales but were uncertain about how to cope with a sudden decline. Like the bar managers whom they so carefully monitored, they too felt themselves to be the fall-guys: "I get fuckin' ulcers from this job," an Area Controller commented on one occasion. "They couldn't pay me for the hours I put in." At another time I was told, "When sales are up, no one says a thing. But if they go down, you can count on hearing about it." In spite of the considerable and expected variation at different seasons of the year, there had been little to worry about in recent times. "It keeps going up—at about 30% a year," summarized one controller. "We don't know where the money is coming from."

The Brewers

The top administrative officers in the Liquor Undertaking saw themselves as beer salesmen. Their target population was urban Africans, but their concern was to sell beer, and any anxieties they felt in their positions related to the machinations of sales administration rather than to the welfare of Africans in any context other than as present and potential consumers. Brewery officials enjoyed a similar clarity of purpose; their task was to provide as much beer of good and consistent quality as the customers could drink ("And with as little alcohol in it as is necessary to satisfy the drinkers," the chief brewer had frequently commented).[5] Their anxieties related to production—they were producing a volume of beer approaching the theoretical capacity of their patchwork but efficient plant. *This Is The Mad House* proclaimed a homemade sign posted over the brewery's production graph, a chart that provided an instant summary of current daily and weekly production and a comparison with the previous year's output for the same period. Their best week in 1970 had been an all-time high of 380,000 gallons delivered; the capacity of the existing plant, a source of unending debate, was estimated at about 400,000 gallons a week. Beer was already being produced in a quantity originally projected for 3 years in advance. Newly completed facilities for some facets of the brewing process were in evidence throughout the sprawling plant, and construction was proceeding on others. "We've had to stop looking for new outlets for beer sales," explained one brewery official. "One week not too long ago they drank up every gallon of beer we brewed. We have to operate this plant 24 hours a day every day of the year in order to keep up."

Higher-ups did not share the view that the brewery was producing at capacity, and the brewers were under constant pressure to increase production. "We're still looking for new outlets," explained one official. "You can't listen to these men on how much they can produce—they can always get more beer when we ask them for it." Staff assigned to outside sales were projecting their market 5 years in advance and talked optimistically of making Bulawayo's beer available "in all four corners of Rhodesia."

The brewery operated with a staff of 14 Europeans and 119 Africans. The African employees included an almost-legendary old man who had worked at the brewery for some 30 years. "He does some tidying up when he's in the mood. Sometimes you find him working late at night. He actually sleeps here. He pretty much lives off the beer, but maybe he goes out for something else to eat once in a while."

[5]E.g., "Kaffir beer keeps the customer alive and healthy by supplying a little alcohol with quite a lot of proteins, starches and unfermented sugars, together with most vitamins of the B complex" (10, *p.* 3).

At Head Office

Administrators at the top of the formal hierarchy of the Housing and Amenities Department felt no uncertainty that theirs was the ultimate responsibility for any problem connected with the Liquor Undertaking. Each individual kept a watchful eye on those above him within the hierarchy and on a number of external audiences and critics, such as the City Council, professionally interested outsiders like the private commercial brewing interests, churchmen, or the YMCA, and organized interest groups like the over 4000 African members of the Municipal Workers' Union, or the 7-man Liquor Undertaking Committee of the African Advisory Board. As public administrators must do in a variety of other contexts, these officials had to mediate among the conflicting demands and expectations of such diverse influences as the profit consciousness of city fathers, their own organization-conscious junior staff, and the best interests of the urban African population as variously perceived by different sectors of both the African and the European communities.

The profit motive, with "profit" taken to include both returning to the city treasury a constantly increasing revenue from gross sales as well as a return on dollars invested, exerted a constant and compelling force on those at the top of the hierarchy. They, in turn, exerted a usually infectious enthusiasm and sense of commitment on those as far below them as assistant bar managers and even upwardly mobile bar hands. While there was no immediate and direct personal aggrandizement involved, it was suggested that the degree of enthusiasm for the Liquor Undertaking evident throughout the administrative ranks was due to a well-nurtured belief that salary improvement for administrative personnel was possible only as long as there was evidence of constantly improving beer revenues. If there were personal convictions running counter to the prevailing enthusiasm for promoting municipal beer, they were not voiced publicly. Indeed, township superintendents and head office personnel typically displayed such enthusiasm that, with few exceptions, they were collectively characterized and caricaturized as having "beer in their bones."

The "Administrative Officer, Liquor" (the "A.O.L.") outranked township superintendents and was directly responsible to the Director and the Deputy Director of the Housing and Amenities Department. In the line-staff organization, the "liquor side" ranked equally with the administrative officer for township administration. Township superintendents were responsible to the A.O.L. for problems related to liquor. The fact that the position of the A.O.L. was more closely related to public administration than to African administration was evident in the recent appointment to the post of a career administrator who had served as a Town Clerk (the senior professional appointment in town management) but had no immediate experience in African administration. The A.O.L.'s job was to oversee the distribution and selling of beer. Although he expressed personal concern for the hundreds of Africans employed in the Liquor Undertaking and was adamant in trying to effect procedures to maintain the quality of both the beer and the surroundings in which it was consumed, he was essentially the vice-president in charge of sales. The graphs which lined one wall of his office gave him a linear picture of just how well he, as well as the municipal beer industry in general, was doing.

In the position of the Director of Housing and Amenities resided the ultimate professional authority for the beer industry, and its professional conscience as well. The Director had accepted his post in 1949, some 21 years earlier, and in recent years it was becoming increasingly difficult for insiders and outsiders alike to distinguish between his role and his personality. To a considerable extent—given the duration of his tenure, the somewhat austere and enigmatic nature of his personality (at least as seen by many of his associates), and the integrity and skill he had brought to the post—the personality of the Department was his personality, its conscience was his conscience, and its sometimes authoritarian, sometimes capricious, and usually carefully considered decision-making process, reflected his own personal style of administering. The townships had developed largely under his direction, and the problems which they in turn had created plagued and worried him. Problems associated with the fantastically successful liquor venture contributed a proportionate or perhaps even disproportionate share of his worries. As "The Chief" worried, so did a number of people under him.[6] His attitudes and

[6]For example, fledgling administrators among the Housing and Amenities Department's junior staff, pursuing formal studies in order to qualify for promotion, sharpened their administrative teeth by preparing practice assignments demonstrating their ability to deal with the issues that plagued their seniors. Although certification for the successful completion of the studies was awarded

concerns were anticipated, pondered, translated, embellished, and generalized upon, to become the unwritten guidelines and preferences attributed to a "They" who constituted someone and everyone at Head Office.

"They" were rigorous in protecting and defending the hygienic standards and the quality of the municipally brewed beer. The farther up the hierarchy one went in search of an opinion, the more fervent were the claims made for the superiority of Bulawayo's own brew. The Director unabashedly argued the quality of the beer. "Have you compared our beer with Salisbury's?" he once asked an interested official visitor from the capital city. "I would be ashamed of Salisbury's figures if I were a brewer." His exasperation with allegations that beer had been "tampered with" at a beer garden were reflected in the strict policy about staff dismissals.

"They" were equally rigorous in efforts to improve the quality of township life for drinkers and nondrinkers alike. The Director's feeling that beer gardens and bar-lounges should be comfortable and attractive was sometimes visible in the facilities themselves and sometimes evident through comments offered by junior staff (e.g., "This tree is really in the way, but the Director won't let us take it down—he thinks these places should be as natural as possible"). The presence of youth centers, social centers, arts-and-crafts centers, health clinics, libraries, playing fields, swim baths, and so on, the endless neat rows of dwelling units that gave every householder—and his garden—a place in the sun, attested to the fact that efforts were being made to do more than merely "create a generation of beer drinkers."

There was criticism that the municipal beer industry was run too

---

through an institute based in South Africa, instruction and examinations were given locally, and one of the two examiners was the Director himself. The following questions, pertaining specifically to the Liquor Undertaking, were excerpted from a list of suggested study topics: "(1) From time to time influential organizations allege social evils stem from the consumption of African beer and liquor. To what extent is this true and, if so, what are the compensating factors? (2) Discuss the part played by beer in the life of the African population and say to what extent a Municipal Liquor Undertaking can fulfill the needs of African urban communities in this regard. (3) Discuss the trend in Rhodesia and elsewhere of the swing from African beer toward European liquor. What are the causes and what can be done to maintain or increase sales of African beer? (4) Some municipalities have created separate liquor undertaking departments whilst others have retained them as a branch of the African Administration Departments. Discuss."

much like a business, with no regard for the welfare of the Africans. The opposite criticism was also heard—that it was run too little like a business. As one inside observer stated, "I think a comprehensive study should be made by outsiders of whether to separate the Liquor from the Housing and Amenities Department. The Director is an administrator, not a businessman. He's running a huge commercial business as a sideline, and he's not running it according to business standards." The charge was almost inevitable when one considered how big the beer industry had become.[7] Any criticism of the municipal liquor undertaking had to be argued in the face of its incredible success, and yet that very success created ambivalence for those at the top who felt torn between their roles as public administrators acting on behalf of professional commitments to community welfare and their profit-making activities as manufacturers. The Director stood firm in his belief that all facets of urban African life should be coordinated under one administration (i.e., that the Liquor Undertaking not be separated from other phases of African Administration). He also argued that administrative-grade personnel needed to be familiar with all phases of the Department's operations. As a consequence, the constant changing and threat of change in assignments was a further source of stress among administrative appointees.

Whether it was due to commitments to efficient administration, to professional or personal preferences for seeking to analyze problems objectively (the Director had contributed a substantial list of scholarly publications to the literature of social science and public administration), to an elected strategy of maintaining an organization open to inquiry and review, or, most likely, to a combination of all of these, the Department of Housing and Amenities exhibited a research orientation through its own efforts and through sponsoring or encouraging the efforts of others. The questions which "They" were particularly interested in or sensitive to in terms of urban drinking provide some insight into their perspective and anxieties.

In addition to the continual gathering and compiling of sales trend data and a never-ending series of ad-hoc reports on particular facets of the liquor industry,[8] the Department had on occasion launched

[7]It was noted that control of the liquor industry in some newly independent African states to the north had precipitated or greatly complicated internal bids for economic and political power.

[8]"This week we're supposed to find out all the foods the vendors sell in the beer gardens, how much they charge, and how much these women make from

major studies to review its liquor enterprise. Two "Work Study" projects completed during the 1960s (17, 18) drew particular attention to factors affecting demand and to short- and long-term trends that could affect beer sales. Such studies provided an opportunity for reviewing the consequences in revenue loss when careful attention was not paid to general supervision and to specific indicators that the operation of a particular outlet might have gone awry:

> "One area revealed a drop of 17% or 20,000 gallons per 4-week period within 2 months. Such a fall-off amounts to £2,000 a month drop in revenue and should have led to an immediate and thorough investigation. Where trends were investigated the finding was usually 'Manager trouble.' In an outstanding example, all staff were removed from a suspect outlet and within 2 months of the new staff taking over the operation of the outlet, sales increased by 67% or 17,000 gallons per month" (18, *p. 11*).

Work study reports conducted in the Liquor Undertaking included such data as the length of time it took to collect, wash, rinse and stack a beer mug (0.542 standard minutes) and contrasts between the time required to serve a half-gallon customer and a gallon customer (0.24 minutes and 0.28 minutes, respectively). Many technical aspects dealt with in the reports have already been noted. The reports dealt also with sources of customer dissatisfaction pertaining to the beer or to the methods by which it was dispensed. The journalistic style of the reports seems characteristic of government reports the world over; one could not help but appreciate an occasional light note, such as a reference to one antiquated beer outlet which, the writer concluded, "must have been erected soon after the arrival of Rhodes."

Officials had also demonstrated a willingness to cooperate with research efforts originating outside their department and over which they could exercise little control. The study conducted by the Alcoholism Research Unit of the University of Rhodesia was a major step in this regard. The discussions and reactions at Head Office to findings reported from that study proved valuable by providing a setting in which official positions, beliefs and concerns were made quite explicit.

---

vending and from 'other sources,'" reported an Area Controller. "That shows you how well they understand Africans up at Head Office. No African woman would have any idea of how much she makes—and she certainly wouldn't tell you, even if she did. They won't even tell you how many children they have."

Two related findings from the Alcoholism Research Unit's study in Bulawayo (86) seemed to give the Director the most pause; whether the percentage of abstainers among African men in the townships was accurate (abstainers represented 47% of the total male population aged 17 and over), and whether the effective per-capita consumption estimates were reliable. The great number of reported abstainers was not in itself a sensitive issue, but the Director expressed some doubts about the validity of the proportion. "If it *is* correct, that's one thing," he observed, "but it may be an overestimate, in which case we would want to see what problems may be due to the self-reporting nature of the survey, to the sample, and so on." Any change in the proportion of abstainers would, in turn, affect the per-capita consumption estimates. The smaller the number of acknowledged drinkers, the higher the per-capita consumption. Estimates from the research indicated an average annual per-capita consumption among male drinkers of about 190 gallons of African beer and about 30 gallons of European beer. Noting all the possible sources of error in arriving at the per-capita estimates, the Director reiterated a need for careful explanation of how the results were derived. "These figures are useful and helpful to us," he summarized, "but I'm concerned with the interpretation of the report among laymen. Right or wrong, the per cent figures reported will tend to stick."

Officials also reacted favorably to my suggestion for embarking on an independent study. They were interested in having an outsider's point of view and, during the months of fieldwork and writing, expressed their hope of (and "confidence" in) receiving a fair and objective hearing. My queries as to whether there were questions they would be interested to have me attend to produced three major points:

1. *Age.* Is drinking really a problem among the young people and, if so, what might be done about it?
2. *Stabbings.* Why did there seem to be a high incidence of stabbings? "I would think conditions here would minimize it, not make it worse," said the Director, comparing Salisbury and Bulawayo. "There's not so much crowding here, and there isn't as much intoxication. People here drink more African beer rather than the European type which is so much higher in alcohol."
3. *Drunkenness.* Is it a serious problem here, out of proportion to other communities? Could the extent of drinking be termed "excessive"? And if two young African men strip to the waist and engage in a fistfight after they have had a few beers, is that really to be taken as drunk and disorderly behavior?

Throughout the fieldwork, prior to the time when I first made a preliminary copy of my work available to members of the Housing and Amenities Department staff, I had the feeling that my attention was assumed to be directed only toward Africans in an inquiry into African beer drinking. Naturally enough, this had been the focus of previous studies made under the auspices of the Department. To an anthropologist, the problem warranted a broader perspective, including an examination of ways in which Whites, although not dramatically visible in township life, were nonetheless a vital part of the urban African milieu. Of all such groups, it was clearly the administrators themselves who played the pivotal roles in every respect of the municipal beer industry except the actual drinking.

However, I have not meant to give the impression that the Department acted autonomously. It was but one of a number of departments directly responsible to the Bulawayo City Council, and departments competed with each other for resources and for favorable review (and budget allocation) by the elected members of the (all-European) Council. Because membership in the City Council was subject to changes at election time, the Council and their duly appointed permanent staff (Town Clerk, City Treasurer, etc.) provided another amorphous "They" to whom senior administrators in the Housing and Amenities Department paid strict attention. "Ultimately, if it came to the real crunch," the Director recognized, "as civil servants we go the way of the people who pay us."

In the early draft of my study circulated among Housing and Amenities Department officials I suggested that those charged with the formal administration of urban African life were called upon to play conflicting roles in which the profit motive, the whole basis of the economic organization of their society, complicated decisions of how they should act on behalf of their African clientele. I suggested that the official position regarding the liquor undertaking was ambivalent and unclear in that the press for profits conflicted with the overall objective of working for the best interests of urban Africans. I pointed out that in most parts of the world officials charged with "native administration" have been antagonists of liquor interests; in southern Africa the municipal governments had usurped control of a commodity most commonly provided through private enterprise. In so doing, they found themselves faced with objectives not totally reconcilable. Administration is not enhanced under such circumstances.

The administrators argued with my interpretation. "Our position is ambivalent, but it is not unclear," they insisted. "The City Council *faces* the ambivalence and *is* ambivalent, but in general the policy is that the beer we provide is a service, not a business. However, Bulawayo *does* regard the provision of a good beer at the right price a part of their provision of the amenities of life." One did not have to restrict a search for ambivalent or conflicting positions to the provision of beer, the Director pointed out; his Department as well as the City Council in general confronted ambivalence in each domain in which it offered services. He pointed specifically to the difficulty of establishing satisfactory policy about housing development—whether to provide more low-grade units or fewer high-grade ones in face of the always critical problem of urban African housing and the competition for the resources of the City Council.

Administrators in the Department not only "coped" (a favorite Rhodesian term) with ambivalence, they contributed to it by extending participation in decision-making among members of the African community. The Liquor Advisory Board, a subsection of the elected African Advisory Board, served as one vehicle for listening to the community and involving at least a few Africans in the problems of township administration. The existence of the Board meant that there were Africans in the community other than Council employees (who might be concerned about holding their own jobs) to whom one could take complaints about the beer gardens and about drinking. Occasionally the Board took a different point of view from the officials (as, for example, when it decided in favor of allowing the introduction of taped music interspersed with advertising announcements into beer gardens on a brief—and unsuccessful—trial basis). Perhaps more important, the Board included African representatives who were not drinkers and who did not approve of the enthusiastic commitment which they sensed on the part of administrators and city officials.

Summary and Discussion: Recurring Themes in Professional Points of View

In listening to the views expressed by individuals whose professional roles brought them into contact with the municipal beer industry—and subsequently realizing the extent of the diversity within each of the professional groups—I sought to identify themes which kept recurring in the expression of various points of view. The commonality in the themes helped to illustrate "integration" in another and more abstract sense by accommodating differing points of view within the African and European communities. The very divergence of opinion could be and was capitalized upon in a rather subtle "agreement to disagree" expressed positively through expectations that there *should* be different points of view. Thus the recognition that some people within either the White or the Black community approved of the municipal liquor industry and that others did not served as a cohesive rather than a divisive function and, along with a host of other differences of opinion, could be drawn upon in illustration of the fact that the community collectively was concerned about the quality of urban African life and was united in a recognition of the need to "do something" about urban problems.

The six themes identified here are *(1)* Cultural Lament, *(2)* Proceeds to Charity, *(3)* Intemperate Temperance, *(4)* Problem Drinkers and Drinking Problems, *(5)* Symptoms Rather than Causes, and *(6)* "Like a Business." Variations within and between these themes provided many individuals with ritualistic ways of viewing drinking consistent with their professional roles.

1. Cultural Lament. The form and function of contemporary urban African beer drinking in municipally operated beer gardens differed from traditional rural drinking even though it can be traced to its indigenous origins. One heard variations on the theme that if beer and beer drinking could be removed from the contemporary setting and *put back as it was* in traditional African society, with only the adult men drinking (and only on special occasions), there would no longer be any problems. This view was popular with some churchmen, especially those caught between personal beliefs felt to be rooted in divine sanction against drinking and an appreciation, intellectual if not always heartfelt, that "we have to accept other cultures." The view was expressed in all circles, however, and it fitted easily into a society where virtually everyone found satisfaction and a sense of security in harking back to his own conception of the "good old days."

Particularly as it relates to African drinking, I have called this view the "cultural lament." It seems to me to be nostalgic at best and dysfunctional at worst. Urban drinking, as others—e.g., Jellinek (63)—have pointed out, tends to be individualistic and profane. It is continuous with but also different from rural traditional drinking, and nothing stands to be gained by wailing that if only old social controls and old patterns could be revived in the new setting, all the complex problems in an urban setting would simply disappear.

2. Proceeds to Charity. There was an inevitable ambiguity in the position the municipalities took as profit-making manufacturers and, simultaneously, as public administrators charged with looking after the best interests of the urban African. The persistent reminder that beer profits were returned to the African community may have been less convincing than officials wanted to believe, since neither White nor Black critics of the beer industry were swayed by it.

To view the problem from the perspective of the administrators, the reluctance to separate the beer enterprise from the rest of their administrative activities was not difficult to explain. Public administrators rarely have an opportunity to demonstrate tangible evidence of their administrative capabilities, and officials were understandably reluctant to give up their chance to engage in an activity which produced such obvious testimony to the effectiveness of their efforts. Their situation seemed akin to the fund-raising activities of voluntary organizations which become exceedingly aggressive about increasing the gross income of their money-making endeavors from one year to the next but can rationalize that their tactics are warranted because the proceeds all go to charity.

Senior administrators touted the benefits derived from beer profits to a degree that could not help but arouse suspicion about whether they felt anywhere near as much ambivalence as they claimed toward the enterprise. To provide further validation of the benevolent nature of the industry, they supplied beer at cost for occasions within the townships where recognized organizations requested donations to support money-making drives. (Alas—the Liquor Undertaking was so used to collecting money for beer sold, and Africans were so used to seeing them assume that responsibility, that I witnessed one incident that verged on altercation over who would actually collect receipts when the Liquor Undertaking had agreed to turn over the proceeds from a specially set-up beer garden to a township organization. The A.O.L. mitigated the problem by suggesting that the club's treasurer collect the money but turn it over immediately to an accompanying municipal employee for customary handling, an arrangement that proved satisfactory to all.) The City of Salisbury also made public note of a long-time practice of regularly providing free beer to blood donors through the local African hospital.

3. *Intemperate Temperance.* With the exception of the beer drinkers themselves (who were not inclined to insist that beer was everybody's answer), the attitudes about providing and drinking beer held by people in different professional roles seemed to be held rigidly and self-righteously. Those who were most insistent on temperance were, paradoxically, the least temperate in their views. Like personal views about religion or politics, opinions about the benefits of drink are often defended as part of each individual's personal encounter with Truth regardless of the position he takes. I suppose that the very approach taken here of examining social drinking for its integrated aspects has precipitated in the minds of some readers an intuitive "taking of sides" for or against the account, and I would not argue that the social scientist claiming an objective and analytical approach to the study of drinking is any less rigidly attached to his point of view than is the temperance worker in arguing his, except that the former

has a professional commitment for considering all cases rather than looking only at those that support his personal view.

Drinking is a subject of human behavior about which most adults have a point of view. Drinking behavior is also a topic on which people are inclined to have opinions on how *others* do and should behave. Given the caste-like organization of Rhodesian society, it was not surprising that many Whites had opinions about African drinking that expressed unfavorable stereotypes concerning both the extent of the problem and the nature of current trends. A statement by a European missionary from a church which did not prohibit social drinking among its own indigenous population, but which took an absolute stand on the evils of drinking for Africans, illustrates the intemperateness: "There is no sort of moderation. Either they don't drink or they are completely drunk." Yet one finds traces of similar attitudes running through the remarks of individuals in other professional roles as well.

To the extent that final pronouncements about African drinking were often similar, it was interesting to note how often the conclusions were rationalized on grounds appropriate to the professional role of the speaker. Each supported his personal conviction with evidence appropriate to his professional commitment. The partaking of alcoholic beverages seems almost inevitably to lead to broad, almost careless generalization, a problem to which social scientists can be as susceptible as the man-on-the-street. Entire nations are contrasted for alcoholism rates (during my inquiries I was frequently reminded that my own native land was "one of the highest" on drinking). Entire continents are compared for drinking styles, as, for example, anthropologist David Mandelbaum's summary that "the kind of drinking done over large parts of Africa stands in contrast to the drinking patterns used over a large part of Central and South America" (83, *p.* 288).

Attention to group norms and statistical studies seems to have diverted attention from the range in patterns of drinking behavior *within each society*. We usually make allowances for the differences in drinking preferences and patterns we see among our own kinsmen, close friends and social strata; perhaps we fail to concede the possibility of similar variation between culturally different groups.

4. Problem Drinkers and Drinking Problems. There are four terms frequently found in studies of problem drinking, and an inspection of them may help to identify some useful distinctions in professional interests and professional points of view. The terms are *drinkers, drunks, alcoholics* and *alcohol addicts.* Each of the terms signals a host of personal meanings in the way that any one individual categorizes others and, equally important, how he allows others to categorize him. Each term also signals professional interest among certain groups and excludes professional interest on the part of others. A drinker may create a problem for his family or be cumulatively building up to a serious medical problem without being any concern to the police, a bar manager or a clergyman. A "drunk" may provide a problem for the police without posing a professional problem to any other group. An alcoholic or alcohol addict may seek counseling from a patient clergyman or welfare worker before he ever becomes a problem to the police and perhaps long after a doctor has made a futile attempt to help him.

My own interest focuses on patterns of social behavior in regard to normal drinking, and my methodology was heavily biased toward observable behavior—what people did, what people said they did, and what they said they ought to do. One can observe people drinking. One can observe drunken people. Yet even establishing that a person is drunk poses difficult questions in terms of absolute definitions, as courts of law the world over are well aware. One does not observe alcoholism. This study has little new information to offer either to students of alcoholism or to those persons immediately concerned about alcoholism among Rhodesian Africans, other than to emphasize in context the importance which anthropologists give to distinguishing between drunkenness and alcoholism and to repeat the caution against labeling drinking that would be judged excessive in one culture as alcoholism in another culture. As Mandelbaum has observed, "Alcoholism in the sense of abnormal, addictive, pathologically compulsive intake of alcohol is not the same as drunkenness, which can be quite normal culturally, and should not be confused with the standard drinking practices of any society" (83, *p.* 287).

Evidence is slowly accumulating on the issue and incidence of alcoholism among Rhodesia's urban populations. Primary responsibility for this research has been assumed by the Alcoholism Research Unit of the University of Rhodesia. Perhaps the single greatest difficulty in studying alcoholism is in arriving at a satisfactory set of criteria for defining it, especially when working in the cross-cultural setting, and the anthropologist cannot help but wonder whether a sociologically oriented research unit charged with identifying alcoholics would be satisfied with its own performance if it failed to do so. Yet if one defines alcoholism in terms of a social definition which calls attention to persistent and excessive drinking that interferes with an individual's life, one is warranted in concluding that there *are* alcoholics in the townships.

In this section I have reported professional opinions and some of the available research, and in the next I shall provide individual case studies in further support of this conclusion. At the same time that we acknowledge an incidence of alcoholism, the available evidence also suggests that compared with other societies, the incidence is low. Drinking in the White Rhodesian community would provide the student of alcoholism with a far more fertile field of inquiry than drinking in the African community, and Africans of social-science bent might well contemplate why many White Rhodesians who seem to enjoy every advantage are such heavy social drinkers.

The contrast in problem drinking between the African and European is enhanced by making the useful distinctions appearing in some of the literature between alcoholism and alcohol addiction (4, 62). The European alcoholic *has to* have his alcohol. Even if he recognizes his problem, he rejects the obvious answer to it. The urban African drinker seems for the most part to have been spared this frightening ultimatum. Though he would rather not do without his drink, it is still physically possible for him to do so. The urban African appears so far to have been spared the tragedy in which those who begin as enthusiastic consumers of alcohol can ultimately be consumed by it.

5. *Symptoms Rather than Causes.* When problems were identified (e.g., that in spite of its low alcohol content there did appear to be drinkers who could be described as alcoholics among African beer drinkers) or acknowledged (the concern of municipal authorities for the possibility that under-age drinking had become a serious problem in the townships), the focus of attention was usually on social symptoms manifested through beer drinking rather than on the social causes which may have led to it (frustration in employment or educational opportunity; overcrowding; political suppression) or to the total milieu in which the drinking occurred. The possibility must always be recognized that the popularity of urban beer gardens and of urban drinking generally are in fact related to the underlying problems of urban Africans only as consequences, not as causes. Thus the more that certain professional groups such as the police, youth workers, or fundamentalist missionaries were inclined to see drinking as "the problem" of urban Africans, the less likely they were to make any contribution toward resolving it.

6. *"Like a Business."* Closely related to the idea of focusing attention on symptoms rather than causes is the fact of the attention given to making or countering charges that the Liquor Undertaking was—or was not—being run like a (proper) business. The integrative aspect of these charges was that anyone could make them, they were inevitably somewhat true, and they put everyone at common purposes by diverting attention from the nature of the business to looking instead at how businesslike it was being managed.

For example, members of the African Liquor Advisory Board who harbored misgivings about the beer enterprise might find themselves discussing how to make the beer gardens more pleasant, their attention being diverted from production to the allocation of profits. Similarly, administrators sought ways to maximize the City Council's investment in capital outlay by running the brewery as near to capacity as possible, a hard-headed and businesslike approach that kept some of the more influential and business-minded council members satisfied and guaranteed spiraling production figures. Ambivalence about the enterprise was not, after all, any excuse for failing to run it efficiently and to maximize its fiscal contribution.

The Prevalence of Ritual—A Final Note

One of the remarkable facets which became apparent in exploring the points of view of professionals such as those described in this chapter was the fact of how people's roles vis-à-vis both each other and urban drinking seen as an institution seemed to provide a reliable guide for predicting their behavior, their beliefs and the aspects of drinking that they would emphasize because of its relevance to their own role. The wary and concerned took their stand on drinking in terms of roles they had accepted in society; these roles enabled them to perceive problems selectively. There seemed to be no surprises. With rather good accuracy we might have predicted the aspects of urban drinking most likely to be singled out by missionaries, medical doctors, social workers or middle managers like the Area Controllers or bar managers.

Collective comments and criticisms about urban drinking tended to be ritualistic in nature. Each individual's role prescribed his basic stance toward the problems, and his comments dealt with ritual observations and complaints in a manner more closely identified with his colleagues-in-role than in the complexity of the problems at hand. Thus each outsider addressed himself to that part of the problem, or that part of the African, that related to his own sphere of influence. Doctors worried about their patients' livers but had only fleeting information about their patients' lives: I recall one ward tour in which my medical host was honed in so finely to a number of cirrhotic livers among recent hospital admissions that he probably never realized that he failed to address his patients as he guided my attention from one liver to the next. Missionaries expressed concern for the hereafter awaiting their potential converts but seemingly failed to appreciate that drinking beer might provide a welcome respite in the drinker's present circumstances. The police played off the realities of their work against visions of Utopia, presumably ignoring that job opportunities in their chosen line of work might be nonexistent were it to become a reality.

Because the administrators at Head Office were called on most often to present the arguments on behalf of a municipal beer industry, they had become facile at reviewing the accomplishments of the enterprise in such a way as to acknowledge its problems at the same time that they endeavored to build a convincing case that left little doubt of the wisdom of the present arrangement. Similarly they observed other ritual-like behaviors in conducting the business of the Liquor Undertaking in the casual but nonetheless formal procedures and protocol of Head Office. Two important rituals were the exchange of standard jokes about "how the beer was doing" and the regular preparation (and presentation to the Chief) of graphs plotting ever-increasing sales. In like vein the bureaucracy required —or at least encouraged—the practice of ritual in the reports demanded of those within the hierarchy: Area Controllers dutifully noted "African beer fresh and good" before signing their daily reports and forwarding them to Head Office, although I do not recall ever meeting an Area Controller who actually drank African beer.

A bar manager receiving a delivery of fresh African beer was expected to taste it (or see that it was tasted, if he himself was not a drinker) before accepting delivery. Whatever function this taste-check once provided had apparently diminished in importance or had disappeared altogether. That is, bar managers said they *could* refuse a delivery, but I never spoke with one who had done so. Nevertheless, the custom persisted. To speed up the delivery process, however, bar managers were usually asked to give their approval after the beer—in quantities up to 1000 gallons—had been pumped into the beer garden's storage tanks. Ritually the bar manager tasted his beer and nodded his approval—just as the intimidated diner nods his approval to the wine steward after a ritual sampling of an expensive bottle of wine ordered for his table. Ritual acts seemed to be associated with all of the roles of those providing the beer and those onlookers whose professional or personal roles resulted in their taking some stand on issues concerning the municipal Liquor Undertaking.

Part IV

Individual Adaptation to Urban Drinking

11

Outsiders and Insiders

ATTENTION is directed here almost exclusively to urban Africans and to their adaptations to drinking in the urban setting, particularly to drinking in the municipal beer gardens. The emphasis is on identifying the nature and extent of individual variation in behavior and attitudes toward drinking. Some broad categories for sorting the urban population are suggested; I must leave to others the work of conducting the surveys that would be necessary to establish the proportion of the urban population that should be assigned to each.

Although data now available allow only approximation, they do draw attention to the fact that either a great majority of urban Africans do not go to the beer gardens or say they do not go. In either case, more needs to be known. If they do not participate in them, what other institutions are providing the opportunities for social interaction? How effective are the community centers, educational institutions or the churches in this respect?[1] If people do in fact drink more beer or spend more time in the beer garden than they care to report, what new insight does such information give us? These questions go beyond the observations reported here, but they pose tantalizing problems for future research.

As the focus now shifts from examining collective group behavior to exploring the range of individual adaptations regarding alcohol

[1]Anthropologist William Schwab, in a study of urban African life in the nearby community of Gwelo some 15 years earlier, attributed to the churches the same integrating functions which I have identified as characteristic of the beer gardens: "In the location [African sector], religion seems to dominate many aspects of life and, for most of the people, the Christian Church appears to be the stable force around which much of their lives is built. It becomes a place of social interaction, where new ideas and concepts are developed, but where stability still exists. The church, for the African, appears to be one of the major links between his traditional world and the new changing world of the location" (115, *p. 117*). Apparently Schwab encountered a great deal of resistance and stress in the fieldwork setting. He reported that he was unable to visit or study the beer hall (*p. 108*).

use, the applicability of the concept of integrated drinking needs to be reexamined. The concept calls attention to how drinking functions among groups of people and how it facilitates their interactions or validates ceremonial occasions when they feel a need to be drawn closer together. Cultural processes that have utility for groups of people do not necessarily have comparable utility for each individual in the group. Drinking provides a case in point. Drinking can serve integrative functions for a group and still be personally disintegrative for some individuals who belong to the group.

In this chapter a distinction is made between Africans who exhibit one of several different urban life styles: those who report that they never go to the beer gardens, those whose participation in the beer gardens seems to contribute to and be integrated with social meaning in their lives, and those for whom drinking appears to have reached a stage in which it has become personally disintegrative and dysfunctional. Further development of the concept of integrated drinking as a social phenomenon is held in abeyance in developing the consideration of individual adaptation. Consideration of the concept will be resumed in Part V in summarizing urban drinking in terms of collective patterns of behavior.

For purposes of organizing the urban African population into some workable scheme for describing various adaptations toward the beer gardens, I have found it useful to dichotomize between "Outsiders" and "Insiders" in terms of beer-garden patronage. These categories correspond somewhat to "those who don't" and "those who do" frequent the beer gardens, but the distinction is not quite so easily made. Not all drinking in the townships was done in the beer gardens and not everyone in the beer gardens was there to drink. The distinction seems most useful in terms of the varied types of behavior that different individuals expressed in excluding or including beer gardens within the ordinary locus of their personal social sphere. Outsiders generally regarded those who participated socially in the beer gardens as "they." Insiders were inclined to identify with others like themselves who enjoyed their beer or who included visits to the beer gardens in their social rounds.

As pointed out previously, the mere fact of one's presence in a beer garden did not ensure that he was an Insider. Many people were professionally engaged with aspects of beer-garden life and periodically went to beer gardens in the course of fulfilling occupational duties, although in occupying these roles they did not participate socially as drinkers. Europeans, whether the "They" of Head Office or a visiting woman journalist from the Orange Free State, were restricted by law and tradition to the status of Outsiders. Even African barmen and managers, who in their off-duty hours might themselves become Insiders at a beer garden, tended during working hours to regard patrons as "they." All Whites visiting in beer gardens, and any Africans who would have defined themselves personally as Outsiders or who were at the moment occupying formal roles that cast them at least temporarily as Outsiders (e.g., African officials, on-duty bar managers and beer-garden employees) usually remained standing when in the beer gardens. (Bar managers were somewhat an exception, since most of them made a point of extensive social visiting with patrons, but such visits were subject to constant interruption to answer telephone calls or return to the servery to handle administrative detail.) By contrast, Insiders tended to be seated or at least to be engaged in conversations with persons seated. Official departmental business enacted in the beer gardens was inevitably conducted while standing; African sociability, other than brief greetings exchanged in passing, was achieved without the European sense of urgency.

Outsiders to the beer garden are described here in terms of five categories: those completely oblivious to beer gardens; those conscious of them but totally disinterested; those who abstained from drinking on principle; elites and others who did their drinking elsewhere; and those who abstained because of personal preference or circumstance. Insiders are described in three categories: nondrinking patrons; moderate drinkers; and problem drinkers. The epitome of heavy drinking is represented figuratively as the point at which the drunken person is as oblivious to the beer garden in which he lies comatose as the most remote Outsider is oblivious to beer gardens as a setting totally outside the sphere of his own life. By capitalizing on somewhat overlapping interpretations of "obliviousness," it is possible to provide a diagrammatic representation of the range of urban African attitudes toward and participation in beer-garden drinking. In Chart 1, the areas of greatest ambiguity between drinkers and nondrinkers, or Outsiders and Insiders, occur at two points: *(1)* in different meanings implied by the term "oblivious," and *(2)* in the ambiguity expressed by those who abstain from drinking for reasons of personal circumstance and who may or may not frequent the beer gardens but who in any case would not consider themselves as Insiders.

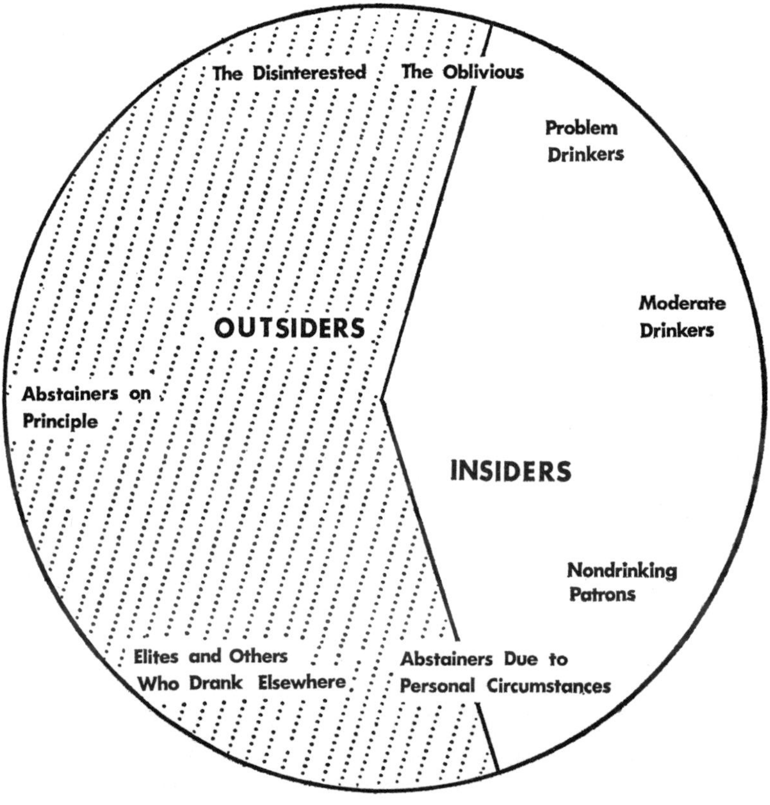

CHART 1.–*Schematic Representation of Adult Urban African Population in Terms of Personal Involvement in Beer-Garden Life*

In Chart 1 opposite points on the circle represent approximately polar positions, abstainers-on-principle holding a position deemed most remote from the moderate drinkers. In the survey of 692 adults 17 years of age and over in the Bulawayo townships (86), some 60% of the sample–47% of the men and 85% of the women–reported that they did not go to the beer gardens.[2] Probably not all of the 40% who included beer gardens among their social haunts frequented them with enough regularity to be considered as Insiders, but no data are at hand for more refined estimates.

[2]May, J. [Personal communication, 1971.] The estimate of 60% is not a simple arithmetic mean, however, since the ratio of males to females aged 17 and over residing in the Bulawayo townships was approximately 1.8 to 1 (1969 Census data).

12

Outsiders To The Beer Gardens

The Oblivious and the Disinterested

COLLECTIVELY, two categories, the Oblivious and the Disinterested, provide a residual group that could be expanded to include virtually anyone, anywhere in the world, who does not know or at least does not care about the presence of the African beer gardens. More specifically, the categories certainly could be expanded to include a majority of the European population in Rhodesia and to those Africans whose lives were not in any way immediately affected by or involved with brewing or consuming municipal beer.

The variation in attitudes about drinking among the African population and the circumstances that provoked it were largely unknown to the White population. "Most Europeans in this town don't have the faintest idea of what goes on this side of Lobengula Street," an administrator in one of the African township offices explained. Lobengula Street served as the boundary marker for the limits of African residence in the direction of the center of the city. If not everyone in the European population was totally oblivious to African township life, it was certainly true that most White Rhodesians had limited and often inaccurate information. They stood generally in awe of the amounts of beer consumed in the townships and of the amount of revenue returned to the city treasury. Their major sources of information about township activities, including beer drinking, were the stories told by their domestic servants (who, by reason of their residence in European areas and their low wages often participated only marginally in township life) and what they read and heard in the news. Newspaper accounts faithfully reported the spectacular aspects of beer-garden life, as the following captions taken from Bulawayo's daily paper during a 4-week period in 1970 attest: "Found stabbed"; "Two deny they killed friend at beerhall"; "Killing was 'act of thuggery': 15 years jail for youth"; "Council to buy 11,000 beer mugs."

How pervasive an influence the beer gardens exerted among the totality of urban Africans was difficult to assess. Although I have suggested the distinction between persons oblivious to the beer gardens and those who could be assumed to be conscious of them but nonetheless remained totally disinterested, I will not belabor that distinction. Obviously many individuals who did not frequent the beer gardens themselves were nonetheless affected because of others who did. The accounts given me provided a full range of behaviors exhibited by the disinterested. Some wives and children worried about father's return from the beer garden and were terrified when he did finally arrive home. Other wives reportedly worried when their husbands did *not* go off to the beer garden to meet their men friends. "My wife doesn't drink," one African explained to me, "but she doesn't care a wink about my going out. If I stay home for 3 or 4 days, she asks me what is wrong. And if she thinks I haven't any money, she'll give me a dollar or two."

My own cook, a middle-aged Ndebele man more at home in the country than in the townships, steered a course completely clear of the beer gardens and found his own social life in the network of full-time domestics living in the various European suburbs of the city. "The beer halls are no good," he said. "The boys are making trouble all the time when they drink in town, and then they kill someone. Drinking may be all right out in the reserves—there's no trouble there—but not in town."

Abstainers on Principle

Some Africans did not participate in beer-garden life (or partake of any alcoholic beverages) as a matter of conscious and deliberate choice. Although such people might regard themselves as "uninvolved," analysis is perhaps better served by thinking of them as a special category of Outsiders. Their position as antagonists of drinking required the existence of drinking behavior in order for them to define what it was they did *not* do. Commitments in this regard have sometimes led to a high degree of organization and proselytizing, both in Rhodesia (e.g., religious groups like the indigenous African churches that maintained part of their identity on the principle of abstinence) and internationally (e.g., political groups like the Woman's Christian Temperance Union, whose frightening messages continue to find their way to the bulletin boards of African schools).

Abstinence from alcoholic beverages shows wide cross-cultural variation. In societies where the partaking of alcohol is moderate or infrequent, the heavy drinker easily stands out as the deviant. But in societies where regular drinking is part of a customary way of life, abstainers are the deviants. Jellinek has noted, albeit in suspiciously sweeping fashion: "French alcohol addicts feel no guilt about their drinking, and thus lack many symptoms which are related to that guilt. . . . The French abstainer does not have such an easy time. In France, drinking is a must; in Italy, it is a matter of choice" (63, *pp. 30, 31*).

I am uncertain whether the role of abstainer is a relatively new one in African societies or whether abstinence has always been accepted (or at least tolerated). My informants argued pro and con, and their firsthand experiences with a generation or two of teetotalers in their own families often conflicted with generalizations made about how "everybody" drank in the old days. One convincing and often repeated argument was that in the old days a man who refused to take beer would have been suspected of poisoning the beer or in some way acting as a witch. Judging from the reported prevalence of beer drinking among all Bantu-speaking peoples, the key role of beer in the performance of rituals validating major events in the life cycle and honoring deceased ancestors, and the fact that one could participate in a beer drink and yet consume rather small amounts of beer from the communal gourd, any adult who practiced total abstinence in traditional life may have been regarded as suspiciously antisocial. Perhaps such behavior was deemed unnecessary among senior adults in whose domain the consumption of beer was seemingly so well integrated.

Contemporary Africans presented a different picture, at least in the way respondents chose to describe their drinking behavior to interviewers during the household survey conducted in Bulawayo.

Men who reported that they were abstainers almost equaled the number who reported that they drank: 218 (53%) reported that they were drinkers, 193 (47%) reported that they were abstainers; and among women, abstainers (81%) outnumbered drinkers about 4 to 1 (86, *p. 7*). Researchers conducting the survey realized too late that respondents may have interpreted the question about drinking to mean public drinking in the beer gardens and bar-lounges and to have refrained from reporting drinking in other contexts, such as social drinking at home, drinking during visits to the country, drinking on ceremonial occasions or drinking only very small amounts.

One also wonders how the frequently complex answers which people give to questions about their behavior are coded into the categories by which data are inevitably summarized and reported. For example, what guidelines could be followed to assure reliable coding between two categories, "Drinker" or "Abstainer," in coding such statements as these two self-descriptions which I obtained during interviews about personal drinking habits?

1. "I don't drink. My father was a Christian. We were never allowed to drink at home. But I naturally don't care much for it. I never spend money buying beer. But I will drink a glass or two with a friend."

2. "Although I come from a Christian home where liquor was not allowed, and I don't drink myself, I sometimes do take sips with my friends. And occasionally I even have one sip too many."

It is hardly surprising that officials in Bulawayo were suspicious of the startlingly large proportion of reported abstainers in the survey, particularly when the Bulawayo study revealed an abstinence rate among men 10% higher there than in Salisbury.[1] Questions were raised only about the proportion of abstainers reported and the unexpected difference between the two communities, however. No one argued that there were *no* abstainers.

Abstainers provide a researcher with the same problem in identifying causes of behavior that drinkers do: reasons for one's idiosyncratic behavior, whether adamant abstinence or hopeless addiction, are complex, highly personal, and at least partly unexamined, operating at preconscious or unconscious thought levels. It is doubtful whether a person can fully explain why he does or does not drink even if he wants to, and from the researcher's point of view, whether or not he "wants to" can be a critical precursor of the kind of interview data provided (46, 80).

For purposes of organizing this account, I have identified four themes for classifying the dominant motives expressed by African abstainers: political, religious, medical, and personal. These categories are not mutually exclusive. Indeed, just as drinkers may defend their "drinking habit" because of a number of benefits which they claim to derive, abstainers are sometimes equally compulsive about the number of valid reasons they can present to support their behavior. My attempts to categorize behavior during an interview

[1]The Salisbury survey found 37% of the men and 80% of the women interviewed claiming to be total abstainers (102, *p. 16*).

were often complicated by the addition of a final concluding statement, "And anyway, I don't particularly care for the taste of the stuff." It does seem useful, however, to distinguish between persons who abstain primarily because of their adherence to a shared principle and those who abstain because of purely idiosyncratic reasons. I discuss the "abstainers on principle" first—those whose abstinence was related to religious conviction and those whose abstinence was related to political motives.

Abstainers for Religious Reasons

No church flatly endorsed drinking. Official dogma and individual interpretation ranged from the perennial plea for moderation through sects which condemned drinking (but not drinkers), to groups which disciplined or even expelled a member for drinking. Few churches exercised moderation in their attitudes toward drinking to the degree they exorcised its temptations.

Religious motives were frequently cited as the basis for individual abstinence, although such reasons were invariably linked to complex systems of personal beliefs and preferences. Among younger persons, drinking behavior often mirrored a parental influence, as suggested by one 18-year-old boy:

> "I have friends who go to the beer gardens. They like drinking beer and to go to meet other women. But I have never even tasted beer. When I was a youngster, my parents belonged to a church—the United Methodist—where they don't allow drinking. And I just grew up with that."

Young people who had begun drinking but who had been raised by parents who abstained primarily for religious reasons, or former abstainers who had become backsliders, observed some personal stress in the compromise between their actual behavior and the expectations others held for them. A 16-year-old boy told me:

> "Each time I go to the beer garden I drink a little more. My mother doesn't know that I even drink beer. I've never told her. She's a Christian at the Salvation Army. Sometimes I go there with her. They don't know that I'm drinking. They describe how bad drinking is. People who don't drink always think it's bad. I don't think it is bad. It's mainly like our food at home, when we're out in the reserves. When you drink beer, you don't feel hungry."

Interviewing in Salisbury for a pilot study conducted by the Alcoholism Research Unit (130) included the following account of a 23-year-old who, after coming to the city, found that he had been unable to sustain a nondrinking habit acquired as a Seventh Day Adventist.[2]

"The informant said that the attitude of his church was that people should never drink alcohol. When he came to work in Salisbury immediately after school, he did not drink, and he stayed for a year or more without drinking. But he discovered that under urban conditions one who went without drinking was a lonely sheep. He found that he was quickly losing many friends who preferred to make new friends rather than come to him, since he would always begin to talk of topics such as the Gospels. He also discovered that the attitude of the bulk of the townspeople was one of derision to anyone who tried to be an ardent church attendant. Sooner or later, he was tempted to follow his old friends, and this meant following them into the pub.

"He said that at the moment he would describe himself as a moderate drinker. He rarely drank during the week, but he would always be in the beer gardens for some hours at the weekend because there was more fun there. He said he still felt heavily about the church but considered it embarrassing to attend services amongst people who now knew he drank.

"His present attitude to beer was still that it was no good for most of the people he saw. He said some people drank and enjoyed themselves at the expense of their families. Others forgot all about God. But he said that he thought drinking was such a big attraction that nothing would be done to stop people indulging in it."

Interviews conducted among drinkers in Bulawayo included persons who claimed membership in churches opposed to drinking. Most such drinkers portrayed a realistic view of their churches' position on drinking, but a few attributed rather liberal views to churches not ordinarily recognized among the liberals. We recorded comments such as "They say we can drink, but don't drink too much" (Lutheran); "They do not mind as long as you do not come to church drunk" (Brethren in Christ); and "At present they are not worried. I think they also drink" (London Missionary Society/United Congregational Church). One steadily employed Shona man who reported that he went to the beer garden every weekday after supper to share

[2]The Salisbury interviewing, totaling 51 interviews, was conducted by a Shona fieldworker, a graduate in sociology. Excerpts drawn from his original field notes appear throughout the remainder of this section. Material reported from this source can be easily distinguished because the interviews were recorded in the third person. The interviews have been quoted with the kind of permission of Professor D. H. Reader, University of Rhodesia.

a gallon of beer with a neighbor, explained that he avoided attending his church (Methodist) in order to keep his name in good standing:

> "They don't allow us to drink. That's why I don't go there. Let's say you haven't gone to church for a year. They start questioning you. And then you tell them about drinking, and they say if you are going to keep on like that they will take off your name."

Abstainers for Political Reasons

In the course of my conversations I identified no African who claimed to abstain for purely political reasons. I would not be surprised if there were such people, but any African so fervently antigovernment as to avoid urban drinking solely on that basis would probably have been active in other political spheres as well, and a decade of purging had driven any inclinations toward subversive political activity far underground. Africans had learned to keep their political opinions—and hopes—almost completely to themselves. In Appendix C, I have provided a statement by one young African expressing a lament about the plight of his people. I had the feeling that many urban Africans shared similar feelings, especially among the younger generation, but no one was voicing such opinions publicly. "Deep down people do feel this way and talk about it among their trusted friends," explained one African man in reviewing political overtones of the kind suggested in Appendix C. "It's dynamite to speak out against the government here," warned another young African. End of conversation.

However, people who were abstainers sometimes included among the stated reasons a criticism of what they considered to be overly enthusiastic support, at both municipal and national levels, for providing the African people with an endless supply of liquor. Antigovernment feelings in the last decade had occasionally been expressed through anti-beer-garden actions. At the height of political reaction in the early 1960s political organizations sent envoys into the beer gardens and demanded that patrons leave their beer and vacate the beer gardens. Mugs were kicked over and people were chased away. Violence was used against those who did not comply, but most people simply ran to escape the onslaught. In subsequent rioting during that period many beer garden facilities were destroyed.

The beer boycotts of that era were part of massive rioting which ultimately proved frustrating to Europeans and Africans alike. To European administrators who had devoted professional careers to providing the level of housing facilities and amenities extant in the townships, the destruction inflicted some deep and lasting psychological scars and a sense of futility toward attempting to "modernize" Africans or to improve their plight. For the Africans, the boycotts and riots provided a review of the power of the Europeans, the growing completely of urban life, and the extent of their dependency on the European economy. Counteracculturative themes were expressed through such collective behavior as taking off one's shoes in the townships because Africans had not worn them before the Whites came. Ultimately the boycotting efforts proved futile, like trying to bring a government post office to its knees by refusing to purchase postage stamps. Even at the height of the rioting, beer service in Bulawayo was interrupted for only 3 days. But the charges made by African politicians had an impact, and the charges were still remembered and aired by nondrinkers, and by many a drinker as well. The following comments by a nondrinker illustrate some political dimensions of abstinence:

> "I don't support the beer halls. I'm quite against them. They are community centers to some extent, but they hold people back. I don't think the motive behind it all is to help people. They are trying to make a man forget his problems. They concentrate instead on the drinking, and on feeling happy. The authorities think it is a social thing for Africans. They don't want us to say that it is disruptive. But I think that it is.
>
> "The government says the beer halls help with welfare. But I think they take it in a false way. They fear that if people are not kept in the beer hall, they might meet together to discuss serious matters which affect the government in other ways."

Another political aspect of the beer gardens was a feeling shared among Africans that instead of serving as places where a man could go to discuss politics, as they had been a decade earlier, beer gardens had become places where one had to be particularly careful *not* tc state political views:

> "You can never discuss politics in the beer halls. You are never sure who is sitting next to you. There are detectives all around, and some private policemen in plain clothes. You get to know some of the people who come to the halls, but you never know everyone. These communities are too big. Especially at the weekends, there are always people you have never seen before, and the beer halls are very crowded. You can go in with your friends, but you can't completely isolate yourselves."

One informant described how during the politically turbulent days of the early 1960s beer gardens had served as political "substations" where minor plots were hatched against nonconforming Africans by members of the major African political parties in the scramble for control of the native vote:

"In those days, people wanted freedom. They talked about it all the time. When a group gathered in the beer garden, they might first crack a few jokes, and then they would get down to the subject of politics. They felt the government was going too slow. Their idea of freedom was to chase the White man out. I was disappointed in the way they went about things, but now as we look back it seems like a Golden Era.

"Things changed with the 'Iron Rule' of the Smith regime. There were a lot of informers around, and the government really worked to clean up the townships. This is still true today, but it is not as severe. The result is that there is not much talk of politics. People just get together to quench their thirst. They are afraid of being arrested and made political prisoners, never knowing when they will be released or who will look after their families."

Abstainers Due to Personal Circumstances

Religious abstainers comprised the largest identifiable category of persons whose abstinence was linked with shared beliefs, and I have suggested that political feelings accounted for or contributed to abstinence based on principle. But there were also bases for abstinence rooted in personal convictions rather than in collective purposes. In the schematic representation (Chart 1) I have indicated that the ranks of those who abstained for individual reasons include both Insiders and Outsiders to beer-garden life. Although the arguments given by abstainers sometimes included a strong personal abhorrence of imbibing, they also extended to such factors as personal health, preferences in taste of beer, or, among juveniles, to strict parental control.

I have already noted that doctors occasionally advised a patient to give up drinking because it was endangering his health and how doctors felt that their African patients often heeded such warnings. Doctors felt that men were more apt to heed their admonitions than were women, the latter sometimes flatly stating a preference to continue to enjoy their drinking at present rather than hedge on the possibility of sustained health at some future time.

Purely personal reasons for abstaining ran a predictable gamut; in illustration I have drawn upon the accounts of two nondrinking informants. The first man reported that he went to the beer gardens "once or twice a month" to visit with friends. He did not drink, and he expressed concern that drinking was becoming a serious problem in the urban community:

"I do not participate in beer drinking. The factors which influence others to drink are not forceful enough on my life. I don't need to drink to associate with others, I don't need to drink in order to quench thirst, and I don't need to drink beer to satisfy my hunger or to forget my troubles. I can do all these without taking beer.

"When I am free and have nothing to entertain me, I sometimes go to the beer gardens to watch tribal dancing and for discussions with my friends who participate in beer drinking. My going to the beer gardens has no effect on my attitude on beer drinking."

The second informant, a man from whom I collected a rather lengthy life history, was more adamant in his disapproval of beer gardens. He never visited them. Initially I attributed his motive in abstaining to a combination of religious and political convictions, until one day he gave me this account:

"My father was a heavy drunkard. I suppose this is why he was always in trouble because of fighting. I remember when I was a young boy he would sometimes come home drunk and start fighting with my mother. Everyone would be panicked when he came home drunk. That made me hate beer drinking. I remember vowing as a young boy that I would never drink, because I would see my father being a happy man and a good father when he was not drunk, but once he got drunk he would start fighting everyone in the house for no good reason. I felt that if I started drinking, I would end up being like him. Then my children would suffer because of my beer drinking. So it is a personal reason, not a religious one, that I don't drink."

Elites and Others who Drank Elsewhere

In the schematic representation of Insiders and Outsiders, "elites" are included among the Outsiders to the beer-garden clientele because those who drank did their drinking in one of several ways that served to maintain them as a group distinct from the ordinary township population. They are represented on the diagram at polar opposites from those oblivious to the beer gardens, since the very essence of being an elite is to be aware of the distinctions and manifest the behaviors that distinguish the few from the many. One such distinction was that urban elites did not show a preference for African beer, typically faulting its low alcohol content—"What's the use of drinking if it has no effect?"

This study did not deal immediately with elites. Sustained focus on beer gardens tended instead to systematically exclude them from consideration. At the same time, certain members of the township community who were especially visible as unique personalities or because of formal roles they occupied were conspicuous by their absence from beer-garden life. Perhaps beer-garden clientele includes individuals from all strata of township dwellers, but essentially the facilities cater to lower-class urbanites.

Bulawayo officials had become acutely conscious about developing attractive "high class" facilities in conjunction with the larger beer gardens. They were conscious also of offering an attractive and more exclusive alternative to the beer gardens through providing bar-lounges. In spite of these efforts, the acknowledged haunt of Bulawayo's African elites was reported to be one of the cocktail lounges operated by an African, that bar having gained in appeal after the closing of an old downtown hotel that had earned a reputation as a community trouble spot because of mixed drinking. As noted earlier, Africans were also served at the city's best hotel.

The author of a comprehensive study of African elites in Salisbury (67) provided me with a brief account of elite drinking in the capital city:[3]

"The drinking habits of this group were fascinating. None went to the public beer gardens in the townships. Some preferred to drink in private with their best griends. Because of their mission backgrounds they felt slightly guilty about their drinking and tried to conceal excesses. Others went only to certain bars. The 'Fed' was very popular, and a group would always be there at certain times in one corner of the room. Some residents said that they did not go to the beer halls because of the type of people that frequented them—the rowdy 'tsotse' type—who liked to fight. Still others preferred to drink in peace at the Ambassador or Jameson. These like to remain at low visibility, e.g., a businessman will drink hard liquor at the back of his store with his close friends.

"Liquor is a subject of joking but there is also some tenseness involved. Since status is tied up with being like Europeans, drinking European liquor represents a move in the 'right' direction."

[3]Kileff, C. [Personal communication, 1971.]

13

Insiders at the Beer Garden

THE INSIDERS, as previously defined, are those who did partake of the beer-garden life. These patrons and their associates will be discussed in categories related to the extent of their imbibing rather than to the time they spent in the beer gardens. These categories are: nondrinking patrons, moderate drinkers, and those whose drinking posed problems for themselves or for others.

The Nondrinkers

Being a beer drinker was not prerequisite for making a casual trip to a beer garden. As a matter of fact, there were virtually no prerequisites at all for going to a beer garden other than being African. The beer-garden clientele included infants and young children accompanying their mothers, youngsters who walked endlessly through the gardens along the main throughways or to make purchases at the dry canteens, teen-agers who loitered about in order to watch the crowds or to engage in some illicit drinking of their own, persons who sat patiently waiting for someone to offer beer, and a variety of other nondrinkers who consumed the atmosphere without consuming the beer. The attraction of the crowds, and special attractions like the live modern music at the bar-lounges, seemed particularly engaging to teen-agers who typically had neither the means nor the expectation of finding so lively a source of public entertainment anywhere else. An unemployed teen-ager who claimed neither to like beer nor to approve of teen-age drinking was a regular music customer: "I go to the beer halls to listen to the bands they provide. Normally they bar children, but they relax it for listening. Usually I go to hear music on the weekends, but sometimes they have touring bands from South Africa on other nights."

Another young man, 19 years old, unemployed, and recently arrived in Bulawayo from the country, noted: "I haven't found any place besides the beer halls where you can go and hear or play music."

A statement quoted earlier described the drinking pattern of a man who occasionally went to a beer garden to visit with friends and who reported that although he did not particularly care for beer, and he never spent money buying it, he did sometimes accept beer offered by others. The Salisbury interviews included an opposite but also familiar case: a man who reported that he spent about $2 a week in the beer gardens, although he did not drink. His purchases were made on behalf of beer-drinking cohorts; he personally had chosen, in his words, to "follow in the footsteps of the Lord." We found a somewhat similar case in our interviews with beer-garden patrons in Bulawayo. The interviewee was a young man employed in a beer garden as a bar hand. He was an abstainer consistent with his religious conviction (Brethren in Christ) but he reported buying an occasional beer for friends.

The Moderates

Excerpts from the extensive interviewing conducted at the beginning of the Alcoholism Research Unit's study in Salisbury (130) make possible a description of a wide range of individual drinking purposes and patterns among those drinkers whom I here refer to as the moderates:

Drinking to feel happy. (Informant: a 40-year-old married man employed as a night watchman.) Mr. M. C. said that as far as he was concerned, beer drinking was not having any bad effects on him. He had never had trouble because of rents or the school fees of his children because of drinking. In fact, he found that drinking made him feel happier, and thought it would be sad without beer. He said that the provision of beer was a good thing. It was easier to get beer nearby for most of the day, and one was always assured of company in the pub, so that the claims of being bored were very little.

Varied reasons for drinking. (Informant: a 32-year-old married man employed as a translator–editor for an African publisher.) This informant reported that he drank more European-type beer earlier in the month and more African-type beer later in the month. He noted that the reasons for which he visited the pubs had varied with time. There had been a time when loneliness had sent him to the pub. After he married, he was no longer lonely and yet he felt he should go to the pub every so often. The mere desire for beer also sent him to the pub. The pub was also an important meeting place where people discussed current issues. Also, there was not very much else an African could do besides go to the pub to drink. The beer places were the nearest recreational places which everyone could reach with least effort. At first he said he had tried to look for other sources of recreation, but on account of distance they proved both inconvenient and expensive. He said he was finding it difficult to find any suitable recreation for his wife who did not drink and who did not work. He said that the recreational facilities available for women did not distinguish according to their educational standard and this made them largely unexciting for his wife.

Women and beer. (Informant: mid-30s, married, a heavy-machinery operator who described a job which entailed spending long periods away from home.) This man said beer and women were inseparable. The places where one found beer, one could easily get women, and vice versa. He said there were occasions when he was forced to get a woman for the night simply because he happened to meet one and was "compelled" to spend money entertaining her with drinks. There were also occasions when he had gone to the pub specifically to look for a night partner but got so drunk that he failed in his original aim and went without one. This was because once he met the woman, she drove him to drink more and more, and when he was thoroughly drunk, other people who were less drunk stole the woman away. In any case, he said, prostitutes were always inclined to take advantage of one's drunken state to maximize their earnings. By running away from a man who had provided her with all the beer she wanted, the prostitute assured herself of a cash payment from the next man.

Beer as a respite from fatigue. (Informant: mid-40s, a factory worker who worked various shifts and who lived in a worker's hostel because his wife did not stay with him in the city.) Mr. A. M. said that he found it too boring to stay in his room in the hostel because it was dark and cold. He preferred to sit in the sunshine, and the pub provided a good place because other things were found there. He said that beer always made him feel brighter when he was on night shift. It removed all his fatigue. He said also that even when he worked through the day he felt he slept soundly after taking some beer. He said he usually drank with a group of friends, but no one in his group could be called a drunkard. On Saturdays they usually got moderately drunk. It was not safe to get too drunk when one lived in the hostels because people there were rough.

A lot of fun. (Informant: S.N., an 18-year-old who assisted his mother by obtaining fruits and vegetables for her to sell.) S. said he had been drinking for 2 years. After gathering the things his mother required early in the morning, he often visited the pub. His reasons were that after a morning of running about he needed refreshment and most of his friends tended to be in the pubs, especially in the afternoons. He said it looked to him like everyone his age who had left school drank, and some who still went to school also drank.

S. said there was a lot of fun to be found in the beer gardens with the boys. They liked to watch how people behaved when drunk. They were also people who were good at telling all sorts of stories, varying from fighting, daring, thieving, and sexual experiences. He said that the most popular people with the boys of his age were those who talked of their sexual experiences. He said that there were certain older people whom they always tried to sit near because of their wider experience and the advice they offered in their talking. S. said that he and his friends talked about what they heard from these older people. There were two or three of his friends who had considerable experience with sex and every so often they would tell of the latest event. The other things they talked about were the latest films they had seen, the latest records in the hit parade, the fiction comics, and about football [soccer] teams, football matches, and individual players.

The excerpts presented above are from drinkers who described their participation in beer-garden life in terms of its positive attributes: companionship, respite, excitement, and the pleasure of the beer itself. Perhaps the most important characteristic shared among these drinkers is that all of them were regularly employed—their presence in the beer gardens was a matter of choice among leisure-time pursuits. Unemployed drinkers, especially those present during working hours on weekdays, seemed more inclined to account for their presence in the beer gardens as an escape from frustrations and anxieties that otherwise plagued them rather than in terms of something they particularly enjoyed about the beer or their beer-drinking cohorts.

The plight of these weekday patrons, particularly the growing number of the "educated unemployed," represents one of the most serious problems for Rhodesia's urban Africans. In our interviews with male patrons in Bulawayo, one out of every three persons interviewed was unemployed and stated that he was looking for work. Collectively this was a young (median age 23 years), basically educated (median 8 years of schooling), predominantly unmarried group of men each of whom had attached himself temporarily to the household of some better-established urban kinsman—thus forming the most common variant of what Stopforth (127) describes as the "enlarged family" so familiar in the townships—while maintaining his primary identity with his own rural family. As a group they spent an appreciable amount of time in beer gardens each week (median 19 hours) and perhaps a larger amount of money (median $1.30) than one would have expected in the face of their financial difficulties. Their reasons for going to the beer gardens are bound up with the limited alternatives they find in their urban lives, as excerpts from three interviews in Salisbury illustrate:

An escape from being lonely and depressed. (Informant: K.M., 25 years old, unemployed for a long period of time until just prior to the interview, when he had taken a temporary teaching position.) The thought of his complete dependence on his mother was one of the things that drove him to form the habit of drinking. He had managed to begin a course at the University but had not done well and had to leave. He had tried hard to look for employment, going to many interviews which required minimum qualifications lower than his own. During those interviews he noticed that he was always sent to be interviewed by African personnel officers while his European colleagues, nearly all with lower qualifications than his, were interviewed by European personnel officers. At the end, the latter were always taken. After 3 months of the same pattern, he stopped going to look for jobs. After that he experienced the most boring period he had ever had in his life. His mother went to work, and so did his elder brother and younger sister. After 7:30 AM he would be in the house alone and lonely. Everyone seemed to have been watching that he was a loafer. He felt depressed when he remained home alone, and he did not like the idea of neighbors looking at him throughout the day and talking about his rotting education. He tried many things. At first he would go to the newspaper office to read free papers. When he had begun to run very short of money for bus fare, he tried to have long, aimless walks. But this did not remove his depressed thoughts.

The only place where he got some comfort was in the pub. He found many friends there. They did not drink particularly much, but there were many advantages. In the first place, there was no embarrassment because the majority of people who came to the pub during working hours were unemployed. Secondly, there was always a chance of hearing where employment chances might exist. And there were games, such as draughts, so that one did not feel every moment of the day ticking away.

Keeping out of trouble. (Informant: unemployed middle-aged man living with friends in the "Old Bricks" section in Salisbury which, the interviewer noted, was "one of the most crowded and unhealthy places of all the African townships.") The informant was bitter. He said he had lost direction in his life affairs when he was jailed and then restricted for about 5 years for political offenses. He pointed bitterly to the fact that he and many others had been reduced to a point of being social rejects. He said he did not have an uncontrollable desire for beer (although he reported spending most of the day in the pubs and said he got drunk whenever he could), but there simply was nothing else to choose from. He said that the present situation was so geared that beer drinking was the only thing he could do without getting into trouble. He said that the life he was now living of going round the pubs, sometimes without money, did not make him happy. But without the drinking, he would feel the effects of his present position in life and his frustrations more painfully.

Beer as food. (Informant: 22-year-old, single, employed as a clerk.) The informant said that after he had left school and before he had started work, he went to the pub largely because he was hungry. At those times, he said, he let others drink the thinner part of the beer at the top and he would drink the heavier part at the bottom of the mug. He said that even now that he had a job he visited the pub at least three times a week plus the weekend. African beer remained his favorite because it was food as well as drink. Usually he did not go home to have his supper, because it was difficult to cook after work, and he found that he often lost his appetite. Beer was the only thing for which he never lost his appetite. During the weekend, he usually drank European beer because he had time to prepare food.

The frustration of not being able to find a job, and the almost inevitable migration to the beer halls as a consequence, represented a rather predictable course of social events. As a Salisbury informant observed, "The unemployment has reduced some people to such a hopeless state that they are resigned to fate; beer drinking has become their only occupation."

Among at least one major African occupational group—teachers, of whom 75% were men—employment seemed to exacerbate drinking behavior. I do not mean to imply that all teachers were drinkers, but rather that male teachers constituted a highly visible occupational status among beer-garden patrons, an articulate, well-educated group, many of whom seemed unhappy with their lot and whose rather heavy drinking may have been related to that unhappiness. One source of the unhappiness was that many persons in teaching posts, particularly those in the lower-paying positions in the primary schools, were not there out of choice. Teaching is not everyone's "cup of tea," but teaching provided one of the few occupations with some promise of status and salary which bright young people could attain solely on the basis of academic achievement. A young man explained, "I didn't want to be a teacher, but there wasn't any other way I could go. Education provides about the only way we can go up." Those who were able to survive the competition for the coveted opportunity of continuing their studies in secondary schools and teachers' college and were then fortunate to find classroom positions often found that life as a teacher failed to sustain its earlier promises.

The following account of teacher drinking, one of several such interviews recorded in the Salisbury study, presents the viewpoint of a teacher in the upper primary grades who described himself as a moderate beer drinker but who was concerned about the excessive drinking of some of his teacher friends and associates.

Teacher drinking. (Informant: a regularly-employed teacher in his mid-30s, married and a family man.) Mr. W. said he had noticed that a lot of teachers at his school went away to drink on the afternoon of payday, and that most continued going away for the afternoon and evening for about the first week after payday. Some of the teachers became broke within the week, and they may go hungry for the last part of the month, or they go around borrowing from the more responsible. A certain proportion of teachers drink merely because they wanted to, he said, and their drinking is usually less heavy and is constant from the beginning to the end of the month. But a large number of teachers in the African schools were frustrated by occupational and political factors, and these factors have produced uncertainty and insecurity among the teachers which has resulted in many of them developing heavy drinking habits.

Occupational frustration grew out of salary disparities, replacement of experienced teachers with those better qualified, and stagnation. Many male teachers who have been in the teaching field are worried by the lack of advancement they are facing. Mr. W. said that although teachers, like many other Africans, feel some amount of political frustration, the type of political frustration which they feel sharply is within their field of service. For example, he said, teachers are not happy to be used by the Government as a propaganda tool. There are many things they would have complained about concerning their services. He said he thought there was more discontent in the African teaching field at primary level than most people knew about, and the worst thing about it is that no one can talk about it.

Problem Drinkers and Drinking Problems

In spite of cautions they expressed about the reliability of such estimates on the basis of field survey techniques, researchers with the Alcoholism Research Unit felt that they had identified a relatively low proportion of "probable alcoholics," persons defined according to guidelines proposed by Marconi et al. (85) as those exhibiting drunkenness every day or during 2 or more working days per week. The Salisbury survey (102, *p.* 32) identified eight men (2.2%) and four women (1.7%) as probable alcoholics. The Bulawayo survey (86, *p.* 23) identified an almost identical proportion: eight men (2.0%) and five women (1.8%).

The complexities of identifying actual alcoholics or of trying to arrive at a satisfactory definition of an alcoholic, given the low alcohol content of African beer and the highly social context in which it was drunk, have already been reviewed. Practically speaking, however, there did seem to be people who were regarded as alcoholics by others. Doctors diagnosed cases where excessive drinking appeared to be the underlying medical problem, social workers identified problems that were caused or at least aggravated by drinking, and most bar managers, at least those in the early-opening bars, could identify a handful of regulars whom they considered to be alcoholics.

The following case from the Salisbury interviews, including some extra comments volunteered by the African interviewer, describes a person who bordered on being defined as an alcoholic, although it is evident that the man perceived his drinking behavior as a reward after a lifetime of hard work rather than as a new problem in his senior years.

A "life of drinking" since retirement. (Informant: in his 60s, retired after 30 years of regular employment in federal government civil service and still the owner of a small shop.) Mr. N. said that beer had become his most valuable "food." He talked boastfully about the fact that he had been living a life of drinking since his retirement. He said he drank all varieties of beer or spirits, and that now he was drinking every day because he was resting after many years of work. His children were augmenting his savings.

He has to have a tot or two of brandy every mid-morning to revitalize himself. He thinks the minimum he drinks is about half a dozen pints of beer and a tot. He didn't know the maximum. He said that he is considered by the people within his "clique" to be the heaviest drinker because he drinks regularly throughout the week. Mr. N. did not know his average expenditures per day, per week, or per month. He explained that he has been very generous to people who cannot manage to provide their own beer.

When asked whether beer was good for him, he said he had never had any suffering attributable to beer. Sometimes he is in the beer hall or any drinking place more hours of the day than he is out. But he finds himself quite happy and sound all the time and can check on his business regularly. [The interviewer had added the following note 2 days after the interview: "I visited his shop to find out something about him. There were three employees there. They were talking about him, saying he had lost all sense of direction and was completely finished. They pointed at several empty bottles of brandy." The interviewer summarized, "This informant should be earmarked for the study of actual or likely alcoholics."]

Loss of a job—or of a series of jobs—was another drinking-related problem. The following interview was with a young man whose drinking had already resulted in the loss of a job on two occasions.

Many hazards. (Informant: 28 years old, married, employed as a market research assistant. The interview was held in a beer garden.) He told me he was giving me his view of alcohol consumption "like a burnt child who dreaded fire." He said beer drinking had led him personally into many hazards amongst which were his loss of two good jobs as an insurance filing clerk. He lost both jobs due to coming to the office conspicuously drunk. After this he stayed without employment for nearly 8 months until he got his present job. He said that because of the hardships he had during his period of unemployment, due to the fact that he had not been saving any money because of excessive drinking, he had decided to stop drinking. He regretted his past.

Family disruption was another problem frequently alluded to in connection with beer drinking. One of my Bulawayo informants described his concern for an incipient drinking problem in his own family:

"I have a cousin who is spending too much time at the beer hall with his wife. He's been divorced once, and I think that is what happened to the first wife. We feel if he doesn't stop drinking, this marriage will end the same way. We were talking to him about it yesterday. I was there with another cousin. We didn't go to his house especially to tell him that, but it came up in the conversation. We talked for about 2 hours. I just told him that what he is doing is not any good. We started by saying he was losing weight, his skin was losing color, and that maybe he was suffering from malnutrition or not getting his vitamins. He agreed. He said that when he drinks beer he isn't hungry and doesn't eat. We told him this might lead to breaking up his family.

"Every Friday when he gets paid, he comes right to the beer hall. We told him that a lot of people lose their money there. He said he had never lost his money. Already he has taught his second wife to drink. We even told him to take the beer home and drink it there. Because he isn't having a proper home. It doesn't look nice if he isn't home when people come to visit. In our custom, we can go to visit at any time, at any hour, without having to make an appointment. It's a home when there is someone at home, as compared with a house only. But he says he is not used to staying at home and he has nothing to do there."

In order to provide specific examples of problem drinking in the Bulawayo townships, I asked social workers employed by the Housing and Amenities Department whether they would provide me with case data illustrative of the more serious problems with which they dealt. Excerpts from the reports of two different cases recorded in the files of caseworkers are provided below. Each of these cases was handled by a male African social worker. With the exception of a few specific local references and indigenous customs, I imagine that the case histories are not unlike those reported the world over by social workers in urban settings that include some incidence of alcohol-provoked problems. At the same time, let me caution once again that the intent here is to illustrate the *range* of alcohol-related problems rather than to measure their frequency. As a matter of fact, municipal casework statistics indicated that, compared with standard welfare case problem categories (e.g., matrimonial, hardship, health, neighborhood dispute, juvenile, education, widow, old age, and "other"), cases laid *directly* to alcohol comprised the least frequent category (less than 1%). As the Welfare Officer acknowledged, however, welfare statistics record apparent prime causes, whereas drinking typically occurs as a complication among the problems presented, especially in connection with two common facets of matrimonial cases that together comprised about one-third of all the welfare cases recorded: assault and nonsupport (35, *p. 123*).

CASE ONE: *Family disruption—interpersonal stress.* (Source: Case Work Report from a township social worker, Bulawayo. Entries had been made over a 4-month period of time and the case appeared to be closed.) *Initial entry:* Wife states that she has been at her husband's kraal in the country for 6 months attending to her sick mother-in-law. Whilst out there, her husband never sent her money. She arrived from the country 2 weeks ago. The children told her that during her absence there was a woman who lived with them and their father. This woman slept in the bedroom with their father. She later confronted her husband with the fact that he was committing adultery during her absence. The husband was very aggressive. She asked him to give an account of what he was doing with his pay. He refused to give any explanation but stated that if she had anything she did not like with her husband's behavior, she was free to leave him alone. Wife complained bitterly that there was no food in the house and the husband had no money. She had no doubt in her mind that her husband was drinking his money with the girlfriend and was supporting the girlfriend as well.

First home visit. I found small children. They told me that their mother was out and could not explain where she was.

Second home visit. I made an early morning visit (8:15 AM). The client was home. She was sitting outside enjoying early morning sunshine. When she saw me, she stood up and went inside the house and invited me in. The room was in disorder and looking very dirty. It looked more like a careless bachelor's den than a marriage quarter. She said that after she had been to see us, family life had improved. Disputes were no longer an everyday issue, and her husband was bringing in his pay which had been at least one source of trouble. I asked the client whether it was still essential that I should invite her husband to my office. The answer was, "Why not? Of course, yes!"

Third home visit. I made this visit in order to find out whether the amount of carelessness found in the house was a day-to-day way of life. The situation was no better. The first thing that greeted my eyes was the dishes lying about in the front yard. Dotted around were little jam tins or baked beans empty tins. As I approached the home further along the garden path, I felt I had trampled on something unusual and that was faeces (excrement). This was obviously of children. I was disgusted about this and I had to move towards a water tap to wash my shoes. She apologized for this and started cleaning around and looking around in the front yard for anything of the sort. I asked how they were getting on. She said things were fine. Her husband was not going to the beer garden. There had been an occasion when he invited her to the beer garden and she had refused. She does not want to cultivate a beer-drinking habit. She said that she was not indicating to me that her husband had totally reformed. She was aware that her husband was trying to deceive her so that she will conclude that all is well and as soon as she returns to the country, the husband will resort to his bad behavior. She knows her husband gambles, drinks a lot of beer, buys beer for some women, and is very fond of making love to spare women in the townships. She has made up her mind never to go away from him.

Wife told me her husband gets $10.00 a week and she does not see the reason why they should starve. He also has some spare-time jobs. He repairs shoes, and he practices as an herbalist, specialising in children when they are sick. But his money is never with him. He drinks it all, to the detriment of his children and wife.

Fourth visit—Office. Wife visited the office and stated that husband is not sleeping in the house. There was no food in the house. He states that he has to work and that this is the reason he does not come home. Husband was sent a letter inviting him to the office.

Fifth Visit—Office. Wife visited the office and stated that husband had refused to come and see us after he received our letter. Ever since he saw our letter, he has been buying food and sleeping at the house. Wife was advised to see us when the situation begins to worsen.

CASE TWO: *Family disruption—individual disorientation and degeneration in interpersonal relationships.* (Source: Case Work Report from a township social worker, Bulawayo. The social worker estimated the couple involved here to be in their mid-50s. Entries had been made over a period of 14 months, the most recent within 2 weeks of my receiving a copy of the report.) The wife visited the office and complained that the previous night her husband had come home late and drunk. He brought a girlfriend and was going to share the bed with this girlfriend in the home. She fought the girlfriend and husband and refused to let them come in. Her husband thereafter abused her and told her that the girlfriend was pregnant by him and he was going to marry her and divorce his wife. The husband then left their home with the girlfriend to go to the township where the girl was staying. She also stated that besides committing adultery, the husband, although employed, was no longer supporting her. He gave her no money for food, and the clothes she had were bought for her by her married daughter by a previous marriage. She stated that her husband drinks beer heavily and unscrupulously.

The wife looked drunk when she visited my office. She talked a great deal. Her skin was peeling off. Her face was dark. I was left without a doubt in my mind that she drank beer daily, that she never ate food, and as a result she was suffering from malnutrition and was affected alcohol-wise in her mind. I asked if she had been to the medical doctors about her state of health. She had not been, but she knew she was sick. She would rather consult the African herbalists than the European doctors because she knew her husband was bewitching her to kill her in order to marry another woman. At this stage, she violently contradicted the idea that her husband's girlfriend was made pregnant by her husband. She knew her husband was not fertile to make a woman pregnant. The witch-doctors had told them this when they consulted them why they could not get children. Her husband had never been married before and had never had children. He was being cheated by the woman so that he would maintain her.

Second visit. Wife visited the office and stated that her husband had not been home in the 5 days since her first visit. She was starving and the rent had not been paid. She knew where he was with his girlfriend. She will go there and smash the window panes and fight the husband and girlfriend.

I visited the house and there was no food. Wife stated that she was eating from the neighbour. I had an interview with the neighbour. He stated that the trouble with the couple was that although they were elderly persons, they drink very heavily. The wife leaves the house in the morning and comes back at night, dead drunk. He has heard the husband complain that the wife has not cooked, cleaned the house, and washed his clothes. The husband comes home late, drunk. Weekends are drinking sprees for both. They go for days without cooking. He finds them to be beyond listening to advice. The couple stay alone in a two-bedroomed house. The husband is employed as a bus driver.

Third visit. One week later. Husband visited the office on invitation. He denied the story told by his wife. He accused the wife of being useless and alcoholic. Although he drinks, to him he was better than his wife. He maintained that he gives his wife money, but the wife spends it on drink instead of buying food and clothing.

Fourth visit. Second month. I held a joint interview with the couple following an arranged appointment. I made this appointment deliberately for an afternoon. I was told that the husband had approached his employers for time off to attend to his family affairs. They both came in and were dead drunk. This interview was one of the most hopeless ones ever made with a family in my experience. They would not listen to each other but cross-accused each other of drunkenness and unfaithfulness. I had to send them out and advised them against the practice of fighting and making noise for their neighbours. I indicated to them that the Housing Authorities were very concerned about family life and if they choose to live the kind of family life they lead, they would be to blame if drastic steps were taken.

Fifth visit. Seventh month, a Monday afternoon. Husband visited the office without an invitation from me. He appeared very angry and threatened violence to his wife. He claimed he came home from work that morning and found his wife dead drunk and asleep with the doors wide open. He had come in at 10 AM. He found a white doek [headscarf] on the bed and in one of the corners of it there was wrapped the sum of 25¢. He woke the wife up and she could not manage to sit upright and talk. He left her sleeping until 1 PM. He asked where she had got money from, and she replied that the money had been given to her by a boyfriend. He was not going to tolerate this but would divorce his wife.

Sixth visit. One and one-half hours later. Both husband and wife in. Wife denied ever telling the husband that she was given 25¢ by boyfriend. The money had come from her married daughter. Husband leaves her without food and she had appealed to her daughter to buy food. Her daughter had given her $1.00. Has he not seen the mealie-meal in the house, or the 10¢ beer on the cupboard? The 25¢ was the balance from the $1.00. The husband seemed to have been very hasty on this matter and wrongfully presumptuous. However, husband remained stubborn and unconvinced by the explanation from his wife. When the couple went out of my office, they became very abusive, shouting and yelling at each other. I had to ask an assistant to escort them home.

Seventh visit. Two weeks later. Husband visited the office and collected the doek and the 25¢ he had brought in when he accused his wife about adultery. He was convinced that the wife probably was telling the truth.

Eighth visit. Fourteenth month. Husband visited the office and invited me to come to his house to see what had been done. We found the wife fast asleep at 3 PM. I could tell the wife had been drinking heavily, so much so that her appearance at the time gave one the impression of a drunk person. She woke up as soon as we entered the next room. All the doors leading outside and between the bedroom and sitting room were wide open. Husband showed me a badly smashed radio and window panes. The house appeared clean. It also appeared she had been out watering the garden. I gained the impression that she was asleep just to take a nap when husband showed the damaged property alleging that she did it because she was drunk. She quickly intervened and said that the window panes were broken because husband was attempting to smash her head on the window panes and instead he missed and pushed his hands on the window panes, hence the injuries and cuts. He broke other window panes in anger because she had closed herself in the house when her husband continued to wish to assault her. She smashed the radio the other day because husband came home late, drunk, and turned the radio full blast to disturb her sleep. She got up and smashed it. Husband assaulted her for that and she was feeling pain. She threatened that if husband continues to do this, whilst he comes home dead drunk, she will one day get a chopper and axe his head off. Husband took a serious view of this and suggested he would notify the police and would not sleep in the house. When he said that he would not sleep in the house, she said he should go and join his girlfriend and leave her alone.

Ninth visit. Fifteenth month. Wife visited the office stating that husband breaks household property and then blames her for all this, yet he does this when he is drunk. She stated that she has been advised by a doctor at Mpilo Hospital to leave drinking alone, because she was peeling and her feet were swelling up. Her feet were swollen up all along the legs. Her arms were peeling off the old skin, showing a new one.

She said she started drinking in 1946 when she was in Johannesburg as a housegirl. She alleges that she drinks because her husband does not give her money for food. She drinks African beer more than anything else. She will also drink European beer or liquor. When she was advised by the doctor, she did not feel well. She had the craving. She found a substitute for craving for beer by buying milk and converting it to *amasi,* a thick milk which becomes sour. She eats this with bread and now she does not crave beer. She thinks that her husband is a heavy drinker. He finishes 20¢ worth of beer [a "gallon" mug] in a short space of time. He keeps beer near his bed until he falls asleep and when he gets up early in the morning he drinks. This is a daily thing, and he spends all his money on it.

This case was presented in detail because it illustrates many of the facets related to serious drinking problems that have been explored separately throughout these pages: frequent and heavy drinking; sexual liaisons growing out of beer-garden associations; accusations, threats, aggravated assault, and destruction of property following periods of drinking; family disruption, lack of individual responsibility, and desertion; improper diet and related medical symptoms; influence of persisting native beliefs on personal decisions; and the patient attention of a trained social worker whose services were made possible by beer-garden profits.

This case, and others like it, ignores one other dimension that should be made explicit; the concern which such cases engendered among many Africans who personally enjoyed the positive attributes of drinking but who questioned whether the apparent determination of township authorities to serve the constantly increasing demand for beer was also serving the best interests of their people. I have noted that there were individuals whose abstinence may have been related to their political views. The following excerpts from a Salisbury interview illustrate the view of those urban Africans who participated in drinking but nonetheless expressed concern for the social and political implications of contemporary drinking patterns.

Drowning in beer. (Informant: a politically active man in his mid-50s who preferred to be "strictly anonymous.") This informant said that since the coming into power of the Rhodesian Front Party there has been a deliberate effort not only to break political thinking by force but also to try and reduce the African's thinking by drowning him in beer. This, he claimed, could be seen easily by the efforts the authorities have made, be it municipal or national, in making the African drinking places the best recreational place of all social facilities presently provided.

He said that the policy of giving the pubs a "new look" by building beautiful brick shades all around the inside of the beer garden, the provision of music, the provision of fire during the cold season, and the provision of snack bars, were all part of this new-look policy which has come into force during the past 6 years. He said this was paying dividends as far as the authorities were concerned, for not only were the African people drowning in beer—both the old and the young—but the authorities collected handsome profits both in money and in political apathy.

He said that the soporific effects of alcohol were apparent in the African, for beer had become the commonest talk of the day over and above other important topics. Many families, especially in the lower income groups, were suffering because the father enjoyed himself at the expense of everyone else in the family. He said that general drunkenness was on the ascendance, and this had begun to affect young people whose future might be in real danger. He said that it was very unhumanitarian for the authorities to keep on increasing the supply of beer on the excuse that the demand kept increasing, when they knew alcohol was a substance whose consumption demanded further consumption. Whatever might be said to justify the ever-increasing supply of beer, especially in the urban areas, honest people must speak out against the bad effects it is having, directly and indirectly, on the social life of the African people.

In one form or another, such allegations have always been leveled at the municipalities engaged in beer production in southern Africa. Similar charges have been made by indigenous groups in other parts of the world.[1] I found no evidence of such a "plot," nor do I actually have any idea of how one would go about either proving or disproving such a general accusation. At the end of my stay in Bulawayo I felt that the comment made in this regard by a person whom I considered to be an astute observer of contemporary Rhodesia was even more apropos than I had realized on first hearing: "I have absolutely no evidence of a conscious design on anyone's part to keep the African population in a perpetual stupor . . . but there are people who wouldn't be above trying to do so."

Yet the very fact that drinking was seen as a potential threat to group life and that its effects were disintegrative for some individuals served also in a socially integrative way to focus urban African thoughts on their own problems and even to engage some members

[1]From a study of alcohol consumption among the Mapuche Indians of South Central Chile, for example, Lomnitz quotes one native informant who aired the classical complaint, "The Spaniard introduced drink in order to have the Mapuche where he wanted him and take his lands away" (79, *p. 294*). The same complaint was voiced to me among the Kwakiutl, and in that setting I was usually included in the accused group of "you guys" (i.e., Whites) blamed for the introduction of liquor and thus responsible for its many unfortunate consequences.

of the White community to help them achieve levels of social consciousness relevant for their urban setting. Although liquor outlets were widely distributed throughout the townships, neighborhood groups in some areas had successfully resisted proposals for specific locations and were beginning to feel that they could exert influence over other matters relating to urban drinking. If trends toward heavier drinking do meet the gloomy expectations of some, it is possible that elements within the townships which give the appearance of being so widely disparate because of differences in language, place of origin, and degree of urbanization will become more fully integrated in coming to grips with new problem aspects of urban African drinking.

14

Becoming a Beer Drinker

THERE is no specific point in time at which a nondrinker becomes a drinker. One of my Shona informants mentioned how mothers sometimes mixed "beer and sadza" to give to a very young child, and one might argue that at the moment a child first consumes such a mixture he takes his most important step in acquiring a taste for beer and for the custom of drinking it in the company of his own people.[1] On the other hand, one of my nondrinking informants, a young man in his mid-20s, described how his own father had assiduously avoided drinking throughout his life and then suddenly, some 4 years earlier, had begun drinking European beer and taking up after women (or perhaps it was the other way around, observed the son) in a totally new pattern of behavior which eventuated in the father leaving his family for another woman.

The fact that a person can become a beer drinker at almost any time (or, conversely, that he can give up drinking at any time) does not preclude the presence of a modal pattern of behavior, and it is an exploration of that pattern with which this chapter deals.[2] To delimit the topic, attention is directed to learning to become the kind

[1]Early induction into drinking has been noted as one of the social and cultural factors contributing to drinking problems elsewhere among Bantu-speaking peoples. Commenting on the "alarmingly high incidence of excessive drinking" in contemporary Uganda, psychiatrist G. A. German has written (38, *p. 21*): "Exposure to alcohol consumption as a pleasant and basic social activity may commence at a very early age. In Teso district of Uganda a beer is brewed from finger-millet (marwa) which is consumed via straws from a large communal bowl. This bowl is the focal point of most social activities in the home and in the village community, just as is the open domestic fire in colder climes. Together with parents and other relatives, the toddler participates in the social group talking and drinking around the bowl, and thus from a very early and impressionable age comes to regard drinking as a warm and pleasant activity inseparable from social behaviour."

[2]Learning to become a drinker has also been described in other cross-cultural settings. See, for example, Honigmann and Honigmann (55), Simmons (120), Ullman (128).

of beer drinker who frequents the beer gardens, with particular attention to the behavior of young men.

One African observer suggested that sophisticated African youths in urban areas today, unlike their country cousins, may acquire their initial taste for beer through the drinking of European-type beers and then subsequently acquire a taste for African beer as their social contacts expand while their ability to spend money on drinking remains severely limited. I was not able to explore this possibility, but it would represent another changing pattern in urban drinking, one that could markedly affect the future of the municipal brewing industries.[3] Interview data in Bulawayo showed the switch to (or stated preference for) European beer to be associated with improved financial circumstances,[4] but that would not negate the possibility that the *preference* for European beer is learned quite early among urban youth, perhaps in the course of nonpublic drinking. In the urban setting, such a preference may be acquired before one ever begins drinking at all.

Induction into Drinking

In their treatise on drunken comportment, MacAndrew and Edgerton offer a cogent observation about the learning of drinking—and drunken—behavior (81, *p.* 88): "Over the course of socialization, people learn about drunkenness what their society knows about drunkenness; and, accepting and acting upon the understandings thus imparted to them, they become the living confirmation of their society's teachings." Statements made about drinking by young drinkers not only lend credence to this observation but also suggest that society's lessons are learned early and thoroughly. The following two accounts, both from boys in their teens, not only describe the particular circumstances surrounding each individual's induction into drinking but also provide a number of cues that show how well an entire and appropriate drinking scenario has already been learned.

[3]Commenting on information obtained through a commercial survey conducted in Salisbury in 1970, the General Manager of its Liquor Undertaking Department noted that European beer seemed to have "gained a tremendous foothold with the younger drinkers" (18- to 24-year-old group).

[4]Approximately 62% of those under the median income in the sample of Bulawayo beer-garden patrons preferred African beer, and 59% of those above the median preferred European beer; however, differences of this magnitude are not statistically significant (.05 level, chi-square test).

The first account of induction into drinking was given by an 18-year-old Ndebele boy, the third of five children. At the time of the interview he was out of school and had been unable to find work:

"I first went into one of these beer gardens to drink when I was 9. I went with three older boys. They were 16, 17 and 19. All of us were Ndebele. They went to buy the beer, and then we went to a corner of the pub. We got one mug and drank it, and then they got three more mugs. We got drunk. Then we went to a teen club where a band was playing. We went there at 7 PM and we stayed until 10 PM and then we went home. When I got home, I vomited.

"When I was still in school I used to go to the beer garden some days at the end of the week. When I was about 11, we went there several days in a row during school holidays. Sometimes we begged money from our fathers. My father thought I was buying sweets with it. Now I go on Saturdays and Sundays and sometimes other days. The last time I went was 2 days ago. I go with a friend of mine who was born here in the township. He is 19 years old. ["Is he Ndebele, too?" I asked. "Of course" came the reply.]

"To get drunk, it takes about 6 shillings [60¢, enough for 3 'gallons' of African beer] or even more for the two of us. I like African beer, but sometimes I drink European beer, too. When we get drunk, we don't become shy. Then we can talk easily to the girls or to anybody. We spend the day enjoying it, just feeling fine. Sometimes we meet new girls, and if they drink, we buy them beer. Some girls do drink beer, but some only drink Coca-Cola.

"I spend about 7 shillings a week on beer. My father gives me an allowance of 5 or 6 shillings a week, and one of my brothers gives me a shilling or 25¢. I spend all my money on beer. My father doesn't know I spend it all on beer, but sometimes he sees me drunk. I tell him I'm not drinking. He is a quiet man, and he doesn't say anything.

"My father drinks African beer at home. He sends me or my brother to get a gallon of beer for him. He doesn't give me any beer, but sometimes I drink some of it. My mother doesn't drink. She says, 'This drinking of yours is bad.' I don't answer. I just go away.

"I become drunk on Saturdays and Sundays. When I am drunk I may go to sleep or I may just have a good feeling. I have a lot of friends at the pub. Some of them work and some of them don't. I've never had a job. I've been waiting at the Youth Employment Centre for months.

"The pubs are not dangerous places. I have seen people get stabbed, but I've never seen anyone get killed. The problem seems to be when people get drunk who don't understand each other—like Shona and Ndebele people—and then they fight. Those who don't drink say beer gardens are bad. Those who do drink don't say they're bad. I don't think they are bad. I never get in trouble in the pubs. But sometimes they won't let me in to the cocktail lounge. They say I'm too young.

"There are about four beer gardens that I go to. Sometimes we stay until 7 or 8 at night. Then we go look for girls or go to the bar-lounges. My father expects me to be home by 10 PM. That's when he finishes listening to the radio. If I'm not there, I just slip into the lodger's room [in many houses a separate entrance to one bedroom facilitates renting it] and sleep with my younger brother.

"Sometimes when you drink, you feel satisfied and don't need to eat. I enjoy beer, but when you wake up in the morning it makes you feel like vomiting. If you go back and drink another mug, you feel all right. Or if you drink hot coffee, or eat some meat or soup, you feel better. It depends on how much you drink. Sometimes you don't remember what happened yesterday. Then you drank too much.

"Even if people say that young boys shouldn't drink, they have their secret places. They see old people drinking and then they go and drink, too."

Another boy, a 19-year-old Shona, attributed his drinking to boredom and to frustration with the educational system:

"I never thought I would drink beer in my life, until I left school. But I was just bored. And it was too hard to bear in mind that I couldn't go on. I had a good grade, but I just couldn't get a place. I still have applications in, but so far I've been getting only 'regrets.'

"I had drunk traditional beer, at tribal dances and so forth, that had been given to me by the old men. In Mashonaland, even the young boys are allowed to drink. I had friends who went to the beer gardens. The first time I went I just wanted to see what it was like. I went with four other boys, one 17, one my own age, and two who were 20 or more. We were all Shona; I find it a bit hard to learn Ndebele. Only one of the boys was a close friend. He had enough money for one beer plus some cigarettes. The other two boys were already going to the beer garden, and we met them on the way. I did not drink beer that time. The next time I went, I tried a sip, and it was nice. Then I started to go regularly. If I get drunk, I just go home. My friends help me. But I really don't drink in most cases. Just during the weekends.

"You hear people say, 'If you drink some beer, you relieve yourself of some troubles.' I mostly drink when I have a problem I can't figure out, or when I want to forget. Education is still my problem, and when I want to forget it, I get drunk. Even if I get to 'A level,' it's still 50-50 whether I could go on to university. It's very disappointing. Our education is going down. With such troubles, there's nothing you can do but try to forget.

"Beer can't really solve your troubles. It's O.K. when you start, because you can get drunk with 1 shilling. But later on you need more alcohol, and you can't get drunk with 1 shilling. So in most cases, the beer gardens are bad. The little you earn, you spend there. Sometimes you'll see a person who doesn't even have a shoe, but he's drinking there. Some people go every day, right from work, and they don't go home until the beer hall closes. And it causes people not to think. They just say, 'As long as I'm getting my beer, it's O.K.' But unless something turns up, I'll keep going to the beer halls.' "

Most boys in similar straits told similar tales. Their full-scale induction into the beer gardens was a function of common circumstances: out of school, out of work, bored and frustrated with their lot, and lacking in alternatives where they could spend time without having to spend much, or any, money. They may have drunk beer before; their changed status as "drinkers" derived more from their inclusion of the beer gardens within the sphere of their regular social rounds than from their imbibing. As one boy explained, his earliest drinking, with beer easily obtained during strolls taken in villages bordering the mission school he attended, was primarily for the purpose of joining in the activities of slightly older boys or enabling him to boast of his activities among his cohorts. He also noted that many boys started drinking because of hunger; in the urban setting the chances of finding someone willing to share beer were far better than the chances of an offer of food. Even without money, he said, it was highly unlikely that he would go into the pub and come out without having had beer to drink. One was almost sure to find friends who had beer.

Youths who did not drink often expressed firm opposition to drinking and voiced specific objections resulting from parental or religious instruction. Those who did drink, although they sometimes acknowledged the problems that drinking caused for others, reported no apparent concern about their own drinking unless they had consciously tried to curtail it or had given up drinking altogether after having felt themselves becoming enslaved by the habit. One boy who described making solitary trips to the beer garden reported that the beer not only tasted good but that as it began to make him drunk it made him feel strong and made talking to others easier for him. Another boy contrasted the consequences of drinking African beer with the danger of drinking anything stronger: "I tried drinking spirits once. But I heard from a friend that if you drink them, you have to go to the hospital. They say that drinking spirits can eat your lungs, and I believe it. But African beer can't hurt you."

Another young man recalled switching to "strong stuff" because he did not like the sour taste of African beer—even when he added sugar to it. He started drinking gin from a bottle that his parents kept at home, taking only a tiny bit at a time so they would not discover the loss. Since his parents were infrequent drinkers, the loss was detected nonetheless, and he received a warning about the dangers of drinking spirits: "They said it would be bad for my lungs and it would make my hands shaky."

Because so many adults with whom I discussed drinking mentioned the problem of youthful drinkers, coupled with the fact that my earliest interviewing in the townships was conducted among older teen-agers in the job-seeker group, I found myself accepting the conclusion, before I had systematically explored the dimensions of that problem, that under-age drinking did indeed pose a problem. An opportunity to be more systematic came in setting up the interview schedule to be used in talking to patrons at the beer gardens. Two questions concerning induction into drinking were included, one dealing with the age when the interviewee started drinking beer, the other with learning who had given him his first beer. Excluding the four nondrinkers who turned up in our sample[5] and two drinkers who did not remember when they started, our drinkers reported a range of ages from 9 to 45 years when they were first introduced to beer. The lad who reported drinking when he was 9 said that his earliest drinking had been done at home: "My father used to give beer to me. I'd see him drinking and I would just cry for it." The median age reported for beginning drinking was 21 years.

To the question "Who first gave you beer to drink?" the most frequent response was "I started on my own," a category that accounted for 19 of the 66 answers. In most of these cases interviewees meant that they had obtained beer through their own volition, although they may have done so in the company of others. Three respondents said that they stole their first beer. The answers of the remaining 70% are instructive in suggesting not only the frequency with which certain people play key roles in inducting others into beer drinking but also in showing the range of people who perform in this capacity: 18 of the 66 were first given beer by a friend or schoolmate; 5 by

[5]The presence of four nondrinkers among our interviewees provides an interesting exception to the claim that the availability of beer leads inevitably to drinking, since two of the four were beer-garden employees and a third operated a shoe-shine concession at a beer garden. The fourth nondrinker was an unemployed 21-year-old who reported spending about 5 hours a week in the beer gardens but who drank only soft drinks and met regularly with a group of friends of similar inclination. "I don't plan to start drinking until I get married," he explained.

father or both parents; 5 by a brother; 4 by a grandmother; 3 by an uncle; 2 by a grandfather; 2 by a brother-in-law; 2 by a cousin; 1 by his mother; 1 by his nephew; 1 by another unspecified relative; 2 by a teacher; and 1 by his employer. Thus, the category "friend or schoolmate" was the single largest, but a collective category comprised of all immediate kinsmen accounts for almost 40% of the responses about the initial givers of beer. The role of parents in this regard warrants special attention.

Parental Influence on Drinking

In conversations about their own drinking, African men frequently referred to the circumstances under which they had observed drinking among their seniors. Some recounted specific parental instruction they had received, as illustrated by the following remark made by a man whose father had been a moderate drinker and who prided himself on a similar reputation: "My elderly father advised me, 'Before you drink, my son, have something to eat. Don't drink on an empty stomach.' "

In one beer-garden interview, a 33-year-old Shona employed as a petrol-station attendant (who reported that he now got drunk every time he came to the beer garden) recalled the occasion 8 years earlier when his father had first presented him with beer: "My father was living out in the reserves. He gave me beer and told me, 'You must drink this, you're grown up now.' That was when I first started drinking. I didn't get drunk that time."

Another Bulawayo townsman, a man of Zulu background, noted:

> "When an African child grows up, say around 18, 19 or 20, there is a little ceremony. I say 'little' because there may be no ceremony at all. But if one is held, the father calls the son into the house and he has a talk with him about serious matters pertaining to the family, such as about the land and crops and the sale of cattle. And the son is given beer, which is brought to him by his mother, and he is given a pipe. These things signify that now he is a man and he may smoke and drink as men have always done, for beer drinking is something traditional with the Africans."

Intrigued by these comments, and particularly the latter one suggesting a continuity in the transmission of drinking patterns from father to son, I decided to inquire into this behavior pattern more systematically in future conversations. My first opportunity came during a week spent with an Ndebele family in the bush. I mentioned the "little ceremony" described by my Zulu informant. My host's response was, "Who told you that? What's his *isibongo* [clan name]?" He had assumed—correctly—that my information was not of Ndebele origin. Subsequently it was suggested that by ignoring the tribal affiliation of my respondents to the question about who first gave them beer I may have overlooked an important distinction between groups. A reexamination of individual questionnaires did in fact reveal that Ndebele drinkers were twice as likely to have reported that they began drinking on their own than were Shona drinkers; that all six persons reporting having first received beer from one or both parents were Shona; and, conversely, that the relatively smaller proportion of Ndebele who received their first beer from a member of their family received it only from a grandparent or brother. The total number of respondents of the two major tribal groups who could remember the person who first gave them beer is too small to be significant (28 Shona, 23 Ndebele) but the difference observed is too significant to ignore. Perhaps others can investigate such differences more extensively.

Several accounts presented in these pages have suggested how parental concern exerted a positive influence on the attitudes about drinking of the subsequent generation. By way of contrast, here is a situation in which an 18-year-old Ndebele boy recounted how his father's attempts to stop him from drinking only made him determined to persist at it:

> "I started drinking beer last year. I was with my friends who were drinking, and they told me to come drink. The first day I was drunk, and my father chased me out. He said, 'Who has given you the right to drink?' I went to my friend's house and slept there. When I came home in the morning, my father asked me again. I told him that I was thirsty and there was no water to drink. He told my grandfather, who is a policeman, to hit me. My grandfather put me in handcuffs and told me to lie down and he hit me with a bamboo. Afterwards, I couldn't get up. I was unable even to move. At that time I didn't like beer, but today I still am drinking it. My father thinks now that I don't drink, but when I drink, I just don't come home. My father drinks at the beer garden, but he says that beer is not good for young boys. I agree with him—but I drink."

Rites of Intensification

From accounts offered by young drinkers, it appeared that the initial induction into beer-garden drinking was usually followed by and reinforced through observing certain practices which served to validate one's status as a drinker and to mark significant changes in that status. For example, in a pattern of behavior somewhat comparable to the "first fruit" rites of many societies, beer drinkers occasionally recounted "first check" rites in which the receipt of a first paycheck at a new job was celebrated among one's friends in the beer garden. A 25-year-old teacher interviewed in the Salisbury study described such an occasion in his life:

> M. said that he had grown to like beer so much that he had begun to have a craving wish for the day he would be able to buy the amount of beer he wanted. His day came when he got his present teaching job. He estimated that about one-third of his pay check must have gone for beer. He said he drank until he felt sick. He also had the first opportunity of providing beer for his still-unemployed friends and for those who had been providing beer for him.

Another Salisbury informant, aged 22, described three objectives which he subsequently achieved once he again became a wage earner. The first was to be able to buy enough beer so that he could get drunk. The second was to be able to choose for himself the type of liquor he would drink, for, as he as well as other interviewees noted, "The appearance amongst one's friends drinking European liquor was as good a sign that one had conquered the 'unemployment phase' as appearing in new clothes." The third objective was to be able to attract women, an objective which he felt he could attain if he had money for buying beer or otherwise entertaining them, and which he was most likely to achieve if he himself had drunk enough to overcome his shyness about making the initial contact. An informant from Bulawayo pointed out that women were not in such short supply there and thus Bulawayo men were less inclined to consider purchasing beer as a prerequisite for initiating an encounter with a woman.

Learning the Urban Drinking Pattern

Although he was only 19 years old, the young man[6] whose account appears below to conclude this chapter has already learned an adult male drinking role appropriate to his age, economic circumstances,

[6]Technically an African male of 19 would be referred to as a "boy" in both Black and White Rhodesian society, in the former because of age, and in the latter because of age and race.

and urban setting. He and a group of three Ndebele friends make a regular visit to one of several favorite beer gardens on Saturdays and Sundays. They take turns buying the beer, usually sharing 2 gallons in the course of the afternoon, enough to make them feel good without feeling drunk. They enjoy the company of each other and of other friends and they enjoy watching the singing and dancing. They disapprove of people who cause trouble in the beer gardens.

Unless he becomes discouraged with city life and returns permanently to his parents' home in the reserves, this young man's life as an urban dweller may change very little in the next few years. Eventually, of course, he will be free to stay in the beer garden later than his older brother now allows him to do. A job will probably turn up for him, and in the interim the Youth Employment Centre offers him a supervised place to go where employment information is available (and, incidentally, is one of the welfare services provided for out of beer-garden profits). Perhaps once he is no longer dependent on his brother as his sole source of financial support he will occasionally drink the more expensive European beer popular with many young people. Still more remote in time, he can expect to get married and start a family of his own, although with his anticipated wages it may be more practical to find a girl who will live with him rather than marry him (a "pot wife"), especially if the girl comes from a family that expects a lobolo payment as part of the marriage agreement. In any case, he and a group of friends will probably continue to meet and share beer together in the local beer garden. Indeed, he himself cannot imagine it any other way.

"I stay with an older brother here in one of the African Townships. He is 35 years old, and he has a job as a baker. I have another brother who lives in town. He has a job as a houseboy out in the suburbs, and he usually comes to my brother's place on his afternoon off. I've never had a job. I'm hoping now that I can get a factory job. I don't know how much I would earn—maybe £2/10 a week [$7 U.S.]. The only money I have now is what my brother gives me, which is usually between 2 and 5 shillings a week, but he doesn't give it to me at a regular time.

"I usually spend the money drinking on Saturday and Sunday with three boy friends who are my same age. We go drinking in the afternoon. Either my friends come to my house or I go to theirs, and then we go to the municipal beer garden together. There are about three beer gardens that are my favorites, and we go to one of them or to the main beer hall.

"We take turns buying the beer. One boy buys it for that day. During the afternoon we drink two of the large mugs. We just pass the mug around, and when we are finished with one mug the person buying it goes back to the window and buys another one. We drink about the same amount each Saturday and Sunday afternoon—it's enough to make us feel good, but it doesn't make us drunk. I never drink so much that I feel bad the next day. Anyway, we don't stay too long—I have to be in the house by 6 PM.

"I like to go to the pubs and visit with my friends. When I go there I see other people I know and I talk to them, too. And almost every Saturday or Sunday the people do some singing or dancing, and I like to watch that. The beer gardens can be a real happy place.

"Sometimes my brother drinks beer at the house, and then he might give me some. But when he goes to the beer hall, he goes with his own friends, not with me. I only drink African beer. I've never tried European beer or spirits.

"As long as people can drink in the pubs without making trouble, they are nice places. Some people can drink all day without making trouble. If people start fighting, it's not so good. Sometimes you see women fighting, but the men fight more. They even use stones or bricks when they fight—they pick them up outside the beer halls and chase each other with them. It's good that they have police in the beer halls.

"I don't go drinking during the week. In the mornings I go to the Job Placement Centre. Sometimes I play soccer in the afternoon. Or else I just stay in the house. I keep the house clean, and take care of the garden. I like to listen to the radio when I'm at the house, especially to music and to the school broadcasts. And I sometimes read if friends loan me their books. But I don't get books from the library—I've never been to a library. And sometimes on Sunday I go to church, but not very often. Maybe I go once a month. It doesn't interest me much.

"About my own drinking, I can say that as long as I have friends who drink, then I'll drink too."

Part V

Analysis

15

The Municipal Beer Gardens in Rhodesian Society

"BEER, the Bible and the seven deadly virtues have made our England what she is," Oscar Wilde observed in *The Picture of Dorian Gray*. Elsewhere man has always extolled the pleasure and necessity of wine, women and song. On the Rhodesian front, one frequently heard a simple maxim for maintaining serenity among members of the indigenous population: "Give an African his beer and his women and he'll be happy."

The pages of this study suggest that the formula, although widely advocated in sectors of the White community and not necessarily inimical in the African one, was not so patently simple, its effectiveness not so universal. At their best, however, the beer gardens were remarkably happy, intensely social places. As grateful as I was that an accident of birth assured that I did not have to dwell in the crowded townships of such an otherwise vast continent, I could not escape a feeling of envy that I could never be more than a formal observer in the simple, robust social settings of Bulawayo's beer gardens.

Mandelbaum (83) has reviewed the pervading norms of drinking patterns observed across cultures by characterizing the kinds of differences that might prevail in some heretofore unknown society observing a completely different set of drinking behaviors from those identified in previous studies:

> "If we should find a people in which women must drink more than men, in which drinking must be done alone or in the company of one's mentors and dependents, or in which the upholders of scripture (whether theological or political) are expected to drink more heavily than do others, we should know that we have encountered a society basically different from others so far known" (*p. 286*).

No such "basically different" society was found in Rhodesia. Its urban African drinkers, like their rural forefathers and their contemporary European counterparts, lived in a society where men did most of the drinking, where they drank in the company of their peers, apart from their mentors and dependents, and where they drank in spite of warnings and pleas from the upholders of both theological and political scriptures. This is not to deny that cultural expectations created some unique drinking patterns among urban Africans that characterized them as a group and distinguished them from other groups. Rather, it illustrates that the variation observed among them fell within customary variations in behavior that have been observed in all societies where drinking occurs and has been studied.

Reports on other areas of sub-Saharan Africa indicate that the problems described here are not so unique to one community as they have been portrayed. They represent in microcosm a cross-section of many pan-African problems, both in terms of widespread beer drinking practices (44, 91, 97, 112, 125) and of the contemporary scene more broadly perceived.[1]

Although urban drinking has disruptive dimensions, particularly in terms of individual excesses, I have chosen to examine drinking in terms of its integrative functions within urban African culture and between the African and European sectors of the community. A review of these integrative functions and some not-so-integrative concomitants provides the basis for the discussion in this final chapter.

Functional Aspects

In describing the long tradition in beer brewing and beer drinking in African society, a number of ways that African drinking is "integrated" within contemporary urban life and in terms of continuity with the past have been identified. I have reviewed how the beer gardens function in a variety of ways that serve to integrate urban life in the face of the presence of powerful disruptive forces: differing languages and tribal origins; constant movement into and out of

[1]Thayer Scudder, for example, has identified four trends cautiously described as "probably pan-African," all of which apply among Rhodesia's African population: (*1*) the individualization of ownership at the expense of corporate kinship groupings; (*2*) the growing importance of ego-centered networks based on both kinship and voluntary association; (*3*) the increasing inclusion of practically all African societies in a frame of reference with both rural and urban foci; and (*4*) the desire on the part of an increasingly large minority to migrate permanently to urban centers where life is seen as preferable to a rural existence (116, *p.* 35).

the cities; uncertainty regarding political developments; and great variation in economic status. Summarizing the economic plight of most tribesmen, one of our beer-garden interviewees noted succinctly, "The pockets are small, and the hands are many." Beer gardens serve as central meeting places in which "haves" and "have-nots" who share cultures or circumstances or both join together in facing their burdens. Some talk about them, some ignore them, and some attempt to drown them.

It is the very ambiguity of the beer gardens—the varied meanings they have to their different audiences—that makes them so multifunctional both within and between the two separate societies that comprise modern Rhodesia. To the Europeans, the gardens serve as testimony of efforts to provide amenities for the Africans, and spiraling sales are taken by the Whites as a sign of the success of these efforts and the appreciation of them by Blacks. The beer industry provides employment and salaries for Whites and Blacks and a critical source of funds for providing services. The beer gardens themselves serve purposes almost as varied as their clientele.

Because of their omnipresence, their over-all importance to the total style of urban life, and their remarkable earning power, the municipal beer gardens also represent an issue of critical community concern, and thus they serve as a possible point of focus for dialogue between European and African interests. The Housing and Amenities Department nurtured this potential, and although meetings between its own officials and members of the African community appeared formal and somewhat ritualistic, the meetings illustrate the recognition of shared problems and the willingness of both groups to establish and expand the precedent for talking about them.

Before leaving the subject of cross-cultural communication, let me note that I do not think Rhodesians have made any spectacular breakthrough in this difficult art. Every such dialogue seemed to labor under three handicaps: *(1)* the lingering influence of a colonial era in which every European was deemed a master, every African a child; *(2)* the probability that at least one party in the communication (inevitably the African) was speaking in a second language; and *(3)* the fact that Africans schooled and trained in a British tradition were inclined to adhere to a formal protocol in which everything is done "properly."

I felt that this last dimension, propriety, proved a great source of misunderstanding as Africans seemed time after time to believe they had beaten the European at his own game by using impeccable logic, citing proper authority, or following accepted regulations "to the letter." Europeans of long experience in colonial service or Native Administration lived quite effectively with the formal trappings of their system but by no means attempted to conduct their lives as though formal procedures were necessary or sufficient for every life situation. Their African counterparts thought they perceived a totally ordered European system of behavior in which regulation replaced interpersonal feelings and even good sense. Africans and Europeans attempting to assess common problems often talked past one another, the former arguing how things ought to be, the latter taking a "realistic" view of how they were.

The consequence of these dialogues was that the Europeans remained convinced that Africans were not ready to manage their own affairs, and Africans pondered that no matter how perfectly they played by European rules, they would never be accorded full status as long as Whites predominated. Such dysfunction in cross-cultural communication did not have its genesis in problems associated with the beer gardens, but that context provided one more arena in which communication was often unsatisfactory, other people's behavior or motives poorly understood. For example, a persistent belief among African critics of the beer gardens was that if the officials would simply "enforce regulations," certain problems (e.g., under-age drinking) would disappear. Europeans seemed not to believe that the answer was indeed so simple—and thus gave no indication of renewed effort in this regard.

If among themselves, especially in the rural setting, Rhodesia's Africans are not totally dependent on a cash economy, at least in their dealings with Europeans they have become quite money-conscious. Beer has come increasingly to be associated with money-making. Far from urban centers, European farmers have long recognized the need for an equitable way to allocate the responsibility of beer making or beer selling among rural laborers. In some areas, commercially brewed beer is distributed through local outlets. In others, farm workers take turns at the lucrative task of preparing beer for sale on the weekend. In most cases the profits remain in African hands. Tribal councils are becoming increasingly interested and involved in beer selling as a way of earning money for community improvement, willing to let the Europeans do the brewing and underwrite the costs of setting up low-cost rural serveries while they take a percentage of the profit on the beer sold. Private commercial interests and the municipalities, each suspiciously eyeing the motives of the other, are competing in the development of these markets and, incidentally, further integrating Africans into the economic system of the dominant society.

Economic motives appeared frequently to be at the root of problems over which Europeans and Africans in official capacity found themselves working together for common causes. Administrators in Bulawayo's Housing and Amenities Department or in Salisbury's Liquor Undertaking Department argued the difference between their profit-making motives and the profit-taking motive of the producers of Chibuku, the commercially produced African beer that reportedly far outsold the beer of either of the two largest municipalities, even though Chibuku's sales were restricted to outlets in rural areas and to smaller municipalities that did not manufacture beer. Commercial and municipal breweries competed for rural markets in ambiguously defined periurban areas. Attention to outlying markets helped catapult wary rural Africans into the economic mainstream and occasionally brought a look of surprise to visitors in distant areas who discovered municipally brewed beer being marketed hundreds of miles from the urban center that manufactured it.

It was interesting to discover that there were conditions under which officials in African Administration did actually "draw the line" in the inevitable conflict between profit-making on liquor interests and over-all concern for the best interests of their township clientele. Such a threshhold was reached over the issue of introducing a new product, "Cola Cane," to the inventory of alcoholic drinks available to urban Africans. Two basic issues were involved in this proposed introduction. The first was whether to allow a larger share of the liquor profits to go out of municipal hands, since Cola Cane was manufactured by private interests. Essentially, however, that question was only a matter of degree. The company that produced Cola Cane was the sole distributor for all spirits bottled in Rhodesia and thus already took a profit on sales of spirits in the townships, although spirit sales were negligible. Furthermore, the manufacturer of Cola Cane was a subsidiary of the company that produced all European beer manufactured in Rhodesia for both African and European consumers. European beer was the antagonist of and ultimate threat to the African beer industry.

The second issue was that Cola Cane, a cane-spirit fortified wine

(18.5% alcohol by volume) was regarded by municipal officials as dangerously potent, especially considering the low price of 25¢ a bottle at which it was first introduced (in October 1970) at duty-free rates in a government effort to encourage the development of a Rhodesian wine industry. Consumer response to Cola Cane in the outlying areas where it was first introduced was so great that by its second month on the market it had to be rationed among distributors. Municipal officials steadfastly refused to allow it to be sold within the townships, expressing great concern over making so potent a drink generally available. If administrators could be accused of having "beer in their bones," there was also evidence of a point at which their other commitments to the interests of the African community took precedence.

The attention paid to beer as a food, and to maintaining its health and hygienic properties, also contributed to integrating the Black and White communities in a common concern about the health of the indigenous population. If Europeans were a bit inclined to overemphasize the healthful properties of the beer, they were in good company among Africans who had always regarded beer as food. Indeed, the consistent attention paid to the quality of the ingredients that went into Bulawayo's beer appeared to be widely recognized and appreciated, and there were innuendos that a carefully planned campaign in the capital city to introduce a "new and fortified" beer—along with a price increase—was a concession that the quality and reputation of its Rufaro beer had deteriorated so noticeably that the only hope for improving the image of African beer there lay in introducing a completely new product.

The beer gardens served as a remarkable tribute to the efforts to which authorities charged with African administration had gone to make urban life palatable and even pleasurable for unending waves of urban migrants. As noted, Bulawayo's townships were acclaimed as among the best, or possibly as *the* best, in southern Africa. Although they were ghettos by design, they had features not usually associated with that term. They were planned expressly for accommodating great numbers of formerly rural Africans of low income. They had provisions for space, sunlight, privacy, even individual gardens—amenities not ordinarily associated with ghetto life. Officials boasted of the efforts they had made on behalf of their African population and had statistics to show how fully utilized were the services they provided. To them, the constantly climbing indicators on graphs illustrating beer sales were a vote of confidence for the quality of their beer and the constantly improving environment in which it was consumed. Officials worked to provide and improve public lighting and public parks, proper playfields and proper drainage, trees for shade and benches to sit on. Without any evidence of the plot of which they were sometimes accused, they sought to keep drinking facilities abreast with township development. The guiding policy on beer garden "siting" in Bulawayo was to develop a new outlet for each 700 to 1000 housing units built.

Municipal officials throughout Rhodesia were convinced that their efforts in providing beer made a vital contribution to urban African life. In a preface to one annual report, His Worship, the mayor of Salisbury, noted enthusiastically, "History may well record that the Rufaro Brewery has been an important contributory factor to the level of happiness which we have been able to maintain in recent times" (110, *p. 1*). "In Bulawayo we would not go so far as to make that claim," cautioned the Director of Housing and Amenities, "but we do regard the provision of a good beer at the right price a part of the amenities of life. We certainly would say that without providing the beer our troubles would have been greater."

Dysfunctional Aspects

The beer gardens do not exert only a positive, integrative influence. I did not collect data necessary to quantify the extent of personal disruption associated with urban drinking—that is a proper topic for research of a more sociological bent. Indices like the community survey conducted by the Alcoholism Research Unit or the records kept by municipal welfare workers suggest that the incidence is low. I have, however, documented problems where excessive drinking has caused the disruption of individual and of family life-styles. Whether alcohol, particularly that which is obtained through drinking the low-alcohol-content municipal beer, is really the culprit in such cases is difficult if not impossible to ascertain. The most serious alcohol-related individual case brought to my attention during the year was that of a man admitted to a mental institution because of deteriorating health and socially unresponsive behavior. The man had been committed by relatives who stated that he had consumed virtually nothing but beer in the previous year and had otherwise refused to eat. The psychiatrist who admitted him explained that such cases typically prove to be physiological rather than psychological in that the observed derangement is due largely to nutritional deficiencies rather easily remedied under medical supervision. The prognosis was for early release if an adequate diet supplemented with large doses of vitamins proved effective. But whether the man would continue to eat regularly after his release or would return to his exclusive beer diet was a question that only time could answer. "Vitamin B" was the prescription for the symptoms; underlying causes would remain undiagnosed.

The number of persons required in the manufacture, distribution and sale of over 15 million gallons of beer each year made Bulawayo's municipal beer industry an important urban employer, and the huge sums of money moving through many hands provided a potential source of stress within the organization. In the competitive employment situation of the urban African, the loss of a job was a serious matter. The very threat of that loss appeared to hang as a specter among beer-garden employees who sensed a feeling of helplessness if the finger of accusation should, rightly or wrongly, be pointed at them. Conversely, recurring attempts to "fiddle" the beer on the part of employees had exasperated officials and led to a detailed system of accounting and control that served as a reminder to African employees that they were being monitored and, in the bigger picture of Rhodesian society, to provide a continual review of the subordinate status of Black to White. Differences in cultural orientation were dramatized in the points of view characteristic of the two groups. Africans probably wondered how White officials could really believe that a man's career of service was less important than a few gallons of beer, especially when individual needs were so acute contrasted with the wealth of the municipal treasury, and they most certainly wondered at the absence of concepts like Christian forgiveness and "due process" in staff dismissals. Not improbably they felt betrayed by a feeling generated in the colonial era and still prevalent among employees that a decent European "boss" should "look after" his Africans.

Yet the adulteration of beer continued to be both threat and reality, and it nurtured suspicion between patrons and employees in the beer gardens and between African employees and their European superiors within the bureaucracy. A long tradition of British propriety recognized that if one could not completely trust a man there was little point in keeping him around. The air of self-righteousness about earning proceeds on behalf of the municipality lent further sanction for the official course of no recourse.

The actions of the administrators sometimes generated their own share of suspicion and seemed to warrant other accusations not actually voiced. For example, limited tank storage space at the Bulawayo brewery necessitated that in order to meet the heavy weekend demand, beer had to be fermented in the least possible time. The shorter the time for fermentation, the lower the alcohol content. Reportedly some brews left the brewery with as low an alcohol content as 0.5%, although 3.5 to 4% was the generally announced specification. The low alcohol content was not a secret—as a matter of fact, it was quite consistent with the brewer's personal opinion that the alcohol level of the beer should be kept at the minimum level acceptable to the market—but neither was the low alcohol content a fact widely publicized among the drinkers. As a matter of fact, the brewer would unquestionably have emphasized that the beer continued to ferment during delivery and in storage at the beer gardens, particularly had he been pressed by a group of indignant drinkers. At the same time, the customer had no guarantee of the potency of the beer he drank as long as he chose draught beer, and even if the lower alcohol content was not due to a shortage in ingredients (other than time) it represented capitalizing on the faith that drinkers might well be expected to have in a "benevolent" municipal industry.

The problem of ingredients was even more delicate an issue, and charges were hurled back and forth among municipalities and between them and the commercial brewers, ranging from friendy rivalry to accusations of "charlatanism" and blatant exploitation. Bulawayo's beer enjoyed an excellent reputation which was usually attributed to a combination of the integrity of the Director of the Housing and Amenities Department and the "genius" of the Head Brewer. As an official in another municipality noted, "The Director has always looked after the quality of the beer for his Africans." The very investment made in maintaining the image and quality of Bulawayo's beer served as a point of frustration to the Director who assumed that lack of attention to beer quality in other municipalities could not help but have an adverse effect on his local market as well.

Inevitably, the allocation of beer profits themselves introduced economic and social strains. In terms of the national economy, there was always some question of whether or not municipalities should be the sole beneficiaries of beer profits, just as at the local level there were arguments over whether the Africans should be the sole beneficiaries of them. Narrow legal prescriptions for how beer profits could be spent, closely patterned after legislation obtaining in South Africa, resulted in overattention to some services (the abundance of welfare caseworkers was the example frequently cited by critics) to the apparent neglect of others, and choices of how funds would be spent reflected imported rather than indigenous cultural values. Urban Africans contrasted community needs as they perceived them with the beautiful—but hopelessly middle-class and European—facilities with which they were typically provided. "We need schools. We don't need to watch the children go swimming!"

From a strictly economic point of view, the municipalities' jealous guarding of their liquor monopolies also served to prevent individual members of the African community from the opportunity to compete for a lucrative source of income. Liquor brewing and selling activities are most often left to private enterprise throughout the world. Home brewing is a primitive cottage industry that can provide a dependable source of regular income as well as an essential vehicle for supplementing income or redistributing cash resources in times of crises.[2] The mass production techniques of the municipal breweries, the consistent qualities of their beers, and relentless official efforts to eliminate illicit brewing had virtually eliminated home brewing as a source of urban income. Through the practice of shebeening, however, urban entrepreneurs were successfully encroaching on liquor profits, and I would predict that regardless of how punitive either police or housing authorities become about it, shebeening will continue in proportion to the generally economic depressed circumstances of people in the townships.

Two decades ago Gussman (43) observed in Bulawayo that those engaged in illicit brewing were not the more criminal or unstable members of the community but tended instead to be individuals attempting to "subsidize their wages to subsistence levels or else to maintain the standard of living to which they have been taught to

[2]Harwood, for example, has described how impromptu beer selling during a year of particular food scarcity among the rural Safwa people of Tanganyika functioned to "reduce the disruptive inequalities in wealth and to redistribute the available money to those with no other income" (47, *p.* 5). Richards, writing about the Bemba, has reported how beer brewing, although illicit, was one of the few means of livelihood open to an elderly single woman and that many widows or deserted wives drifted near town for this purpose (104, *p.* 322*n*).

aspire" (*p. 102*). An empathetic African observer painted substantially the same picture of the shebeen queens of the 1970s. From almost every point of view shebeening was disapproved of, he observed. It was disruptive not only to the community but also to the shebeen operator and whoever else comprised her household: "Shebeen homes are not happy homes." Nevertheless, shebeening remained virtually the only income-producing activity for these women whose own lives had become closely linked with the urban setting at the same time that they found themselves without means of support for themselves and their dependents. The police complained that shebeening thwarted the maintenance of law and order because there was no control over shebeens, and between the complaining of neighbors and the frustration of the police it was likely that authorities would continue to deal harshly with infractions, at least among flagrant violators. Officials might pause to examine the socially integrative functions performed by such practices, however, and to consider the limited alternatives available to women in similar circumstances. Although it is understandable why officials have tried to discourage unattached women from remaining in the city by limiting their prospects for obtaining housing, such an approach may be unrealistic in terms of the real forces at work in transforming a rural society into an urban one. Customs like shebeening are cultural solutions to problems perceived by members of the urban African community, just as the establishment and perpetuation of municipal brewing industries have been a response to problems perceived by members of the European one.

The overriding dysfunctional aspect of the municipal beer industry was that, like attention focused on other aspects and services provided by the Department such as housing or welfare services, the attention was drawn to what was happening between Africans and Whites rather than on important human dimensions on which little or no action could be discerned. I would venture that in the Housing and Amenities Department, and in the person of the Director of that Department, Rhodesia's Africans already had an ally with years of service to African urbanization and general improvement for the African people vis-à-vis their not-so-recent intruders. Criticism and concern about closely managed aspects of African life drew attention away from other issues that were outside the domain of the Department—particularly issues relating to African economic development. Rhodesia's urban Africans needed more than the extra dose of vitamins B and C with which Salisbury's "Rufaro Super Draught" was going to be fortified. It was certainly right to see attention given to such details as the nutritive value of the beer or the number of benches in the beer gardens, but it was unnerving to realize how seldom anyone's identification of critical social problems went beyond the mundane. It was not the selection of topics that seemed wrong, but the emphasis which various topics received. The emphasis on accountability and administrative detail tended to stabilize and sanctify the existing status quo. African discontent was funneled into acceptable channels—a man would tell you he would like to see more beer mugs or benches but dared not elaborate on the comparable shortage of wages, jobs, or political representation.

Officials in the Housing and Amenities Department reminded me that the "big" problems were outside their immediate domain. They were right. But what seemed to me to be Rhodesia's single greatest problem from a humanitarian point of view was that everyone disavowed responsibility for major social issues. Few Whites, and virtually no Blacks, were venturing any action that had a hope of changing the status quo. The plight of contemporary Rhodesia seemed to be foreshadowed in an observation made by Edmund Burke almost two centuries earlier: "The only thing necessary for the triumph of evil is for good men to do nothing."

Under such adverse conditions for social change, overattentiveness to providing the "amenities" provided a lot of full-time jobs, a crucial source of income, and a manageable sort of problem for airing African and European concern. It also served a useful function in easing the social consciences of White Rhodesians in terms of their humanitarian instincts and real and imagined criticisms from a skeptical and to some extent hostile world outside. Those who chose to believe that the Black man wanted and preferred to remain solely among his own people were assuaged with evidence they could cite to show that "the Africans have it just as good on their side as we do on ours." These were signs in the wind that even more amenities might be provided in the townships so that Africans would have less need to come into the White community to share its theater, its cinema, its lovely parks, and its seats of higher learning or technical training. Europeans were ever conscious of how easily they could be "overrun" in a country in which they were outnumbered in the ratio of 20 to 1 by an African population increasing so rapidly (3.6 percent according to the National Economic Survey for 1970) that it could conceivably double within 18 years. Such feelings about the desirability of having Lobengula Street continue to function as a boundary for the social activity of the two communities were assuaged by the knowledge—or the assumption—that the African had been provided sufficient amenities on his side and thus had no legitimate claim to come into or remain in the European areas after working hours.

In Conclusion

Persons immediately concerned with the quality of urban African life might benefit from further consideration about the broad categories suggested here for distinguishing Insiders from Outsiders at the beer gardens. Although I can appreciate why officials responsible for municipal brewing are inclined toward market-type or consumer-reaction surveys aimed at their substantial market, I find myself intrigued by the apparent fact that the majority of township dwellers do not participate in beer-garden life. My research became so focused on the beer gardens, and I sought so diligently to ascertain the dimensions of their integral role, that I think I would have argued at one time that no urbanite escaped their influence. Perhaps that claim could be supported, but one would have to stretch the point to do so. How interesting instead to discern the alternate life styles and networks operating, and to study whether these alternatives could be enhanced by even a fraction of the attention and investment of resources that had been directed at the beer drinkers. The Welfare Section had not been remiss in attempting to provide community services—indeed, its history is recorded along a trail of well-intentioned program starts—but its models appear to derive singularly from the Western world. Perhaps Africans say this is what they want—and maybe they even believe it—but more attention could be given to what people actually *do,* and to what choices they make when they are faced with real alternatives, rather than what they *say* when questioned about hypothetical ones.

Along the lines of action research or developmental research, I offer an observation that could conceivably alter the official policies governing beer-garden operations and expand the social functions the gardens fulfill in the townships beyond a mere "liquor undertaking." The point turns on this question: Is it really essential to operate the beer gardens as though they are "pubs"? Need they necessarily be cast in a Western model dictating that drinking in public is a behavior that needs to be confined to a closed "den" reserved essentially for men and from which children are barred by law and gentlewomen by propriety? Such strict age-grading and privacy of public behavior appear to be Western imports. Maybe there are other models—and other labels—by which these social centers could serve an expanded community function to emphasize and capitalize on the integrative aspects of African perceptions about drinking rather than constantly reinforce all the assumptions about drinking which were introduced by the Whites.

Westerners make some peculiar assumptions about drinking that come into perspective when viewed cross-culturally. One assumption noted in Chapter 1 is that unless it is done with extreme moderation, drinking invariably leads to social problems—that it is invariably disruptive. Cross-cultural study contradicts that assumption. Another is that public drinking is different from private drinking and, more particularly, that other people's drinking is different from (read "more of a problem than") one's own. We rigorously insist that young people be kept out of public places where liquor is sold, yet sometimes flaunt drinking behavior in front of them at home.

In African society, there is and has always been a distinction between participation in beer *drinking* and participation in a *beer drink.* Beer drinking was exclusive, but the beer drink as a social occasion was not, and the customary drumming, singing and dancing were activities in which all could join as participants or onlookers. It is apparent that the same activities in the urban setting attract both young and old to the beer gardens, just as they have always attracted attendance at beer drinks in rural areas. One wonders what would happen if more emphasis were put on the "garden," less on the "beer." Suppose, for example, that gardens were operated as parks where beer was available (and could continue to be a feature) but where all the community could congregate, even though the actual sale of beer was restricted to adults. Such a perception could literally force paying more attention to providing a wide range of services (e.g., adjacent play areas for children, supervised recreation for young people, music available for all to enjoy) in addition to beer. Perhaps the Welfare Section could take responsibility for the operation of the park complex, holding jurisdiction over the bar manager whose domain would be less encompassing but would still include control of the "liquor side."

I hardly make this suggestion in order to reduce problems—it might greatly increase them. Monitoring who would be allowed access to beer would be more difficult, and the creative talents of the administration would be hard pressed to find ways to compete with the tradition of the urban beer garden already firmly established. But even the greater visibility of some of the problems of township life might help the search for solutions to them. More public drinking among the under-aged would not necessarily be worse than the same drinking carried out in seclusion. More teen-agers in parks might possibly leave fewer preying on hapless victims in the streets. More people coming to the beer gardens would not necessarily make for more drinkers, but even if it did, it might help emphasize that drinking need not be incompatible with family life in the urban setting if fathers need not disappear behind high walls to join their cohorts.

I do not suggest that Western-type bars be eliminated in the townships. Conceivably an effort could be made to operate more of them, and perhaps certain areas at the open parks would still be set aside for bars and lounges. The very presence of such establishments as closed-off restricted dens for adult drinking could serve to emphasize the distinction between a tavern or cocktail lounge in the best European tradition and a beer drink in the best African one.

I see at least two important reservations which probably expose the naiveté of this suggestion. The first is that the idea of emphasizing or formally encouraging that beer drinking is appropriate for adults but not youth cannot by itself resolve under-age drinking as an urban problem. The element of social control is essential. Sanctions potent in the rural setting become ineffective in the impersonal setting of the city, and external, formal controls will certainly be required. Where such responsibility can most effectively be placed I do not know. This is the same enforcement problem which liquor authorities confront anywhere that an arbitrary age is established as the basis for determining whether one is sufficiently "mature" to purchase intoxicating liquor on his own. In the case of African beer, what would happen if the minimum age was raised and the onus was put on anyone who supplied beer for a younger person? Conversely, what could happen if the age restriction was eliminated? Is age the best, or only, criterion for judging who can drink?

The researchable dimension concerning youthful drinking seems to lie in the realm of learning more about variables related to individual drinking patterns and to enhancing those conditions which appear to contribute to choosing appropriate and nondisruptive alternatives in individual drinking patterns. Enhanced economic opportunity, either directly through employment or indirectly through the eternal hope which contemporary Africans invest in formal education, would appear paramount among the critical variables which humans can influence.

My second reservation is the recognition that the idea of expanding the recreational aspects of the beer gardens is neither new nor novel. In a report written well over a decade earlier under the same administration (quoted in the section reviewing historical aspects of urban drinking) recognition was given to the fact that beer gardens were "much more" than mere drinking places, they were "real recreation centres where people can relax, gossip, have small parties, sing, and dance," an aspect that "should be greatly encouraged." It is apparent that efforts were expended in this regard. In place of the single and decrepit old Main Beer Hall, one found almost two-score clean, well-managed beer gardens and bar-lounges, with present sites constantly being improved and new sites being planned to prevent overcrowding in existing facilities. Yet it is questionable whether the beer gardens had become "much *more* than drinking places." Instead, one sometimes heard the opinion among personnel on the Liquor Side that if more liquor revenue could be channeled back directly into the development of beer garden facilities, they could be made far more attractive for the patrons who were paying for them.

Among the eternal dilemmas faced by local officials (and by public administrators everywhere) was the realization that no matter what they did for their township population, it would never be enough. Efforts to provide anything new or unusual in the townships were sometimes met with such enthusiastic and overwhelming response that the best of intentions at the outset produced highly undesirable consequences. It was a fact of life that any event planned to give recognition to an improvement in township life was also an occasion to bring in a sufficient number of municipal policemen to prevent the activity from being overrun. Guards routinely monitored bar-lounge entrances to maintain the exclusiveness of those places, and both routine and special detachments were on hand to restrict attendance at any special township occasion. Trying to do things for "everyone" had inevitably proved unrewarding, frustrating and impossible, producing the kind of "ambivalence toward the target/

clientele group" characteristic of any administrative organization (132).

A recommendation such as this to "open up the beer gardens" will probably be met with a cool intellectual embrace and subsequently dismissed for its one most glaring shortcoming: impracticality. Yet a thoughtful analysis of the real limitations of the proposal, including some administrative introspection regarding resistance to such an innovation, might result in some modest assessment of whether urban administrators can capture more of the integrative dimensions of African beer drinking by considering alternatives on how drinking is viewed in different societies. What new insight might we gain if we considered the African perspective that drinking is as "natural" as the intake of food and looked at British and American customs of confining such activities to drinking dens as unnatural and socially disruptive because they require so total a commitment from the individual before he can participate?

Returning to problems of broader social science scope, I have an inclination to recommend further inquiry into the lives of nondrinkers as a basis for comparison with drinkers. Attention should also be called to Barry's suggestion (4) that students of alcohol addiction ought to compare excessive drinkers with moderate drinkers rather than abstainers. As Barry points out, "Avoidance of alcohol represents an extreme social control against potential excessive drinking, and comparisons of individuals, ethnic groups and preliterate cultures seem to indicate that excessive drinkers are more differentiated from moderate drinkers than from abstainers" (*p. 466*).

The student of alcohol and culture might find it of interest to analyze relationships between variations in tribal origin and differences observed in drinking behavior in the beer gardens. Were the differences that people thought they perceived real ones? If not, what were the bases for these cultural stereotypings? And if so, how are present-day forms of "drunken comportment" linked with a basic personality structure continuous with one's past heritage? What might a closer examination of differences between Shona and Ndebele drinking customs reveal?

Perhaps most exciting to the student of culture change is the process of cultural diffusion as exemplified in the gradual shift toward the acceptance of and preference for European beer going on right before his very eyes in the urban setting. Who is making the "switch" and who is resisting it? Is the decision purely an economic one, as some insist? That is the argument put forth in Salisbury, by a manager who feels that if the price of the two beers was comparable, "we wouldn't sell a drop of African beer." In Bulawayo such an opinion is taken as further indication that the capital city is in trouble for having paid too little heed to the quality of its own brew.

It is interesting to note the role which officials are playing in expediting or retarding the diffusion of European beer. The point should be emphasized that as preferences for European beer become more marked, the municipalities will gradually be eased out of their role as brewers and will function only as distributors for commercially produced beer. Laying claim to serving a generally "more sophisticated" urban African, Salisbury's liquor outlets already emphasize the availability of European beer to a greater extent than do those of Bulawayo. The success of Bulawayo's *Ingwebu* beer leads officials there to feel considerable optimism about the future. One wonders if those officials have perhaps grown more committed to the product than the consumers have and how such personal commitment might affect change along the traditional–modern continuum.

It is impossible, as I noted at the outset, to ignore the broader political context in which urban African drinking takes place, so I shall conclude with a note in that regard. The segregationist policies of those in power ultimately affected virtually every interaction which a social scientist might be interested in studying and also set the scene for what cannot help but develop as a fascinating chapter in the resolution of the Black–White encounter on the African continent. Future trends in urban African drinking in Rhodesia could provide new insight into the nature of the relationship between alcohol use and other aspects of culture, just as time and changing political conditions will make it possible at a future date to better discern whether, or the extent to which, alcohol has served as a recognized vehicle of social control or whether it is only an incidental factor which assumed increasing influence in the course of Western domination.

I have not intended to create the impression that political frustration was driving people into the beer gardens. Whether urban African drinking was a direct response to political tension and hopelessness or, on the other hand, was virtually independent of the political climate, pursued as a traditionally based social activity, or the extent to which drinking may have fulfilled both functions at once, I was never able to ascertain. Clearly there was a political dimension in *attitudes* about the provision of beer, but how that relates to actual drinking *behavior* remains unclear. The few Africans who rendered an opinion felt that politics and beer were compartmentalized and were regarded independently in people's thinking. That possibility fits nicely with the hypothesis cited earlier and proposed by Lemert (73) that drinking is an institutionalized behavior pattern operating relatively automonously and is only tenuously related to other aspects of culture.

Based on the great variation from nondrinkers to heavy drinkers observed among urban Africans, I think it would be more useful to analyze individual cases rather than to rush to broad and encompassing generalizations, and I call attention to the recent studies conducted by Graves (40, 41, 42) into drinking patterns among North American Indians migrating to the city of Denver. Graves has eschewed cultural explanations (because, he insists, they conceal our ignorance of what the underlying processes actually are) and concentrated on analyzing individual differences. He has hypothesized that migrants who have a personal goal structure least in line with urban opportunities will experience the greatest psychological conflict about remaining in the city and will therefore do the heaviest drinking. There are some important differences between Indian migrants to Denver and African migrants to Bulawayo, particularly in such dimensions as the nature of the rural–urban continuum and the extent of integrated drinking as that concept has been used here, but Graves' attention to individual personalities and circumstances points to a promising line of inquiry and suggests the importance of augmenting survey data with individual case histories.

The world had relentlessly pursued its own form of apartheid in isolating Rhodesia through economic and political sanctions applied in the wake of the Unilateral Declaration of Independence in 1965 and still in effect throughout the period of my fieldwork in 1970–71 and subsequent writing. All of White Rhodesia had been adjudged guilty of a course of action that clearly did not reflect the will of everyone, although it did appear to represent the view of the majority among the essentially White electorate. Officials of the Housing and Amenities Department, and those similarly charged with the responsibility of African administration in other municipalities, worked within a political system which some approved and some did not but all accepted at least to a sufficient degree that they could operate within it. They accepted the fact of the existence of townships and a prevailing climate of opinion holding that Black and White life styles were sufficiently different to warrant spatial segregation. But given these assumptions, they turned their attention at least in part to improving the lot of the urban African and to facilitating the inevitable transition toward Westernization and modernization. I was frequently reminded that African people living in South Africa and Rhodesia had the highest standard of living enjoyed by Africans anywhere on the Continent. Officials pointed to the townships with pride and repeatedly asked me what I thought of them. I could only repeat forthrightly that at the same time that I greatly admired the effort which had been made and the obvious accomplishment achieved, I was glad that I was not required to live in them.

In every culture there is evidence of compassion and concern. If White Rhodesians were somewhat blatant about perpetuating a caste-like social structure inherited from their colonial past, they certainly had not ignored their African countrymen nor failed to contribute significantly to their human condition. The beer gardens and other amenities seemed designed to make urban life tolerable when it might not otherwise have been so. Provision of an ample supply of beer and places in which to drink it did not appear to be a plot to "keep the African down" as much as an attempt to capitalize on an element of past tradition to provide some respite from the present. In the years ahead history will record whether the African continent was amenable to the continuing presence of a White Africa coexisting with a Black one. Practices associated with urban African drinking may little affect the course of that history, but they will most assuredly reflect its outcome.

APPENDIX A: BREWING

Traditional Brewing

The following recipe describes the brewing practices among the Ndebele as reported from several rural areas in Matabeleland. This recipe is the customary "44-gallon" one, the unit of beer most frequently brewed because of the availability of 44-gallon metal drums used for marketing petroleum products.

I. Raw Materials

A. One 4-gallon tin of rapoka (uphoka). Rapoka, or "finger millet," is a small brown grain grown specially for the brewing of beer.

B. Two 4-gallon tins of ground meal. The meal may be ground from maize (corn, or "mealies") or from the coarser kaffir corn (*amabele*).

II. Preparing the Imvubelo or "Yeast." Time: 4–5 Days

A. Soak the rapoka grain to start germination and leave until germination takes place, a process which take 3–5 days. Usually the grain is put in a basket and water is poured over to moisten it. The dampened grain is then covered with grass or skins to keep the seed moist. The germinating corn is called *imtombho.*

B. Spread the imtombho in the sun and leave it until it is completely dry.

C. Grind the dry grain into a powder, *imvubelo.* Traditionally this was done by using either a stamping block or grinding stones. The imvubelo serves as the yeast for the brewing.

D. Rapoka is not always available and is not grown in all areas. In its absence, other grains (e.g., kaffir corn, in double quantity to the amount of rapoka called for) may be used. Whatever grain is used must be germinated, dried and ground in the same manner as the rapoka in order to function as the yeast.

E. Regardless of the combination of grains used in preparing the beer, it is the grain that is to serve as the imvubelo that is ritually addressed by the head of the household on the day before brewing begins, if the beer is being prepared for the *umbuyiso* ceremony. It is the absence of this crucial and sacred step in brewing for the umbuyiso that leaves some doubt in the minds of many urban Africans about using municipal beer. As one of my sources of information on traditional brewing explained, "I don't think the spirits are as fussy as all that. If you can't get the corn to brew beer yourself, you can explain it to them and then use the municipal beer. But *just how effective this is,* I don't know."

III. Preparing the Fermented Paste. Time: 1½ Days

A. Take the two 4-gallon tins of ordinary ground meal and pour it into 22 gallons of boiling water. Stir this mixture slowly until it becomes a thin paste. Let this paste cool.

B. Mix this paste with half of the imvubelo (made from the germinated grain in *II* above) and leave until fermentation takes place—about 1½ days. The fermented mixture is called *inhlama.*

IV. Cooking the Porridge

A. When the inhlama is ready, bring another 22 gallons of water to boil. Pour in the inhlama as prepared in *III* and stir the mixture slowly for 2–3 hours on the fire until it feels ripe and thickens into a brown porridge. Then let the mixture cool. This porridge, *umhiqo,* is edible and, eaten with sugar, is a popular treat for children.

B. To the cooled umhiqo add the remainder of the imvubelo prepared earlier. Leave for a day or two until fermentation takes place, as evidenced by bubbling activity. During fermentation the mixture will thicken.

V. Pressing out the Beer

A. The liquid must now be pressed out of the thick, fermented mixture. The traditional Ndebele method for accomplishing this was to pour the mixture into a basket "strainer" woven out of *ilala,* a variety of palm. The basket could be wound tighter and tighter to squeeze out the liquid beer. The remaining pulp, *insipo,* can be used as animal feed.

B. The beer is put in a calabash or clay pot and is now ready to be served. It can be stored in these containers for about 2 days, but by the 3d day it may start to sour. If it does, it can be mixed back with fresh beer and will make the fresh beer ready more quickly. Usually beer is drunk the day it is ready.

Municipal Brewing of African Beer (Bulawayo)

This account of municipal brewing has been excerpted from a brief explanation prepared by the staff of the Steeldale Brewery in Bulawayo and distributed to interested visitors.[1]

I. Raw Materials

1. Malt. Malt is any corn which has been allowed to germinate to controlled degrees. This process is carried out as follows:

(a) The corn is soaked in water for about 8–14 hours, depending on the type of corn used. This process is called "steeping." *(b)* The water is drained off from the corn and the corn is allowed to germinate. The rate of germination is controlled by varying the temperatures, usually between 23–28 degrees Centigrade. *(c)* Germination is then halted by heating the corns and then drying the corns to the required moisture content. The process is called "kilning." *(d)* The corn is now ready for milling (crushing). The aim of milling is to separate the white substance from the husk and crush it to flour without damaging the husk.

[1]City of Bulawayo (19). Quoted with permission.

Biochemical Changes in the Corn During Malting. The part of the corn which contains the starch is converted to simpler carbohydrates and the enzyme diastase which carries out the above conversion is further developed. The reason why germination is controlled is because excessive germination will result in the carbohydrates of the corn being lost. A very important guide to the quality of the malt is the "diastic power" defined as the ability of malt to convert starches to sugars by diastase enzymes. It is the sugars which are required for fermentation.

Types of Malt Used in the Production of African Beer: (1) barley malt; *(2)* kaffir corn malt; *(3)* rapoka malt; *(4)* inyati malt. The last three are usually blended together.

2. *Maize.* The most commonly used varieties are: *(a)* Dent maize; *(b)* Flint maize. All maize to be used for brewing is degerminated, that is, the embryo of the corn is removed. It is the embryo which contains the fats and oils which, if present in quantities in excess of 1%, will result in a slowing of fermentation with a final product having a poor head and reduced shelf life.

3. *Rice.* This is the ordinary commercial household rice.

4. *Yeast.* Within the yeast cell are the enzymes zymase, hydrolase and maltase, whose function is the transformation of sugars into CO_2 and alcohol. The yeast cell contains 60–70% nitrogenous substances. Yeast also requires a certain supply of energy in the form of sugar, nitrogen, vitamins, e.g., "Bios," oxygen to enable it to undergo normal and efficient growth.

5. *Water.* The composition of the water used for brewing has a pronounced influence on the character of the beer being produced.

The Brewing Process

The following diagram summarizes the brewing process:

STAGE 1: MASHING

Raw Materials Used
Barley Malt
Kaffir Corn/Rapoka Malt
Kaffir Corn/Inyati Malt
Molasses

↓

Lactic Acid Generators

↓

Lactic Acid Pressure Cookers

↘

STAGE 2: COOKING

Raw Materials Used
Grits
Melite
Kaffir Corn Meal
Rice
Kaffir Corn/Rapoka

↙

STAGE 3: FERMENTING

Raw Materials Added
Fermenting Beer or Yeast
Barley Malt
Kaffir Corn/Rapoka Malt
Centrifuging ⟶ *Storage*

1. Mashing. Raw materials used: barley malt; kaffir corn/rapoka malt; kaffir corn/inyati malt; molasses.

All the above raw materials are mixed with water in a mash tun (tank) at varying temperatures and thereafter transferred to the Lactic Acid Generators. It is in the Lactic Acid Generators where the production of lactic acid is carried out. When the lactic acid concentration (degree of acidity) has reached the desired level, the liquid is transferred to the Lactic Acid Pressure Cookers, brought up to the boil and held for 30 minutes to kill off all the lactic acid bacteria and hence stop the production of any further lactic acid. The liquid is thereafter transferred to the fermenters.

The purpose of mashing is to render soluble and to dissolve as much of the valuable contents of the malt as possible. The final products of mashing are maltose, dextrins, pentosans, proteins, mineral substances and tannins. All the above products are due to the enzyme, diastrase, which causes *(a)* liquefication of the starch paste; *(b)* cleavage of the starch to dextrins by the enzyme a-amylase; *(c)* finally the hydrolysis of starch + dextrins to maltose by the enzyme "'b-Amylase."

2. *Cooking.* Raw materials used: maize grits; melite; kaffir corn meal (unmalted); rice; kaffir corn/rapoka malt.

All the above raw materials are mixed with water at a suitable temperature. The temperature is held for 30 minutes. After this stand the temperature is raised to boiling and held at this temperature for 2 hours. After this it is cooled down and transferred to the fermenter.

During the cooking process the following biochemical changes take place: *(a)* gelatinization of the starch molecules, the liquid becoming a thick porridge mass; *(b)* liquefication. The "a" amylase enzyme breaks down the gelatinized starch to a soluble, more liquid form. The "b" amylase enzyme further breaks down the liquid starch producing the sugars maltrose and, finally, glucose.

3. *Fermentation.* We now have collected in the fermenter the prepared raw materials from the Lactic Acid Pressure Cookers and from the cooker. *(a)* The temperature is brought down to a specified temperature and held for a short period. *(b)* The following raw materials are now added: barley malt; kaffir corn/rapoka. *(c)* Cool down further. *(d)* Add a further quantity of kaffir corn/rapoka malt. *(e)* Cool down to room temperature. *(f)* Add fermenting beer (or yeast). Fermenting beer is beer which is already in an active state of fermentation and is added to the fermenter to initiate a quick start to fermentation. *(g)* When the beer in the fermenter is in an active state of fermentation it is transferred through the separators into the storage tanks. The separators are used to remove excess incipo, or pulp, from the beer. The maximum fermentation time in the fermenter is usually 18 hours

Main Changes Taking Place during Fermentation: (a) The valuable contents of the malt added are rendered soluble. *(b)* Due to the presence of the "a" amylase and "b" amylase enzymes the soluble starch extracted from the malt is converted to maltose and sucrose which are in turn hydrolyzed to glucose. *(c)* The glucose required by the yeast will now initiate fermentation producing alcohol plus carbon dioxide, which are both by-products of fermentation.

Three Basic Stages during Fermentation: (a) Lag Phase: The yeast remains in an inactive state absorbing all the nutrient requirements removed from the wort (unfermented beer) such as essential mineral salts, amino acids and "B" group vitamins. *(b) Krausen Stage:* Yeasts, taking in all necessary food requirements, start to multiply by the process of budding, and fermentation starts. *(c) Fermentive Stage:* The yeast reaches its most active state of reproduction and produces alcohol and carbon dioxide. *(d)* The lactic acid bacteria produced in the Lactic Acid Generators attacks the available carbohydrates thereby increasing the acidity of the beer until the characteristic sharp taste of African beer is produced. An important point to bear in mind is that the growth of lactic acid bacteria at high temperatures inhibits the growth of most other harmful beer bacteria.

4. *Specifications of the Draught Beer in Final Product*

Calorific value	=	350–400 calories per pint
Alcohol % v/v	=	± 3.0
Lactic acid %	=	0.30
Sugar (glucose) %	=	± 4.0
Solids (wet) %	=	19–22
pH	=	3.5

Food Properties: A pint of African beer contains carbohydrates, fats, proteins, 0.5% mineral matter, vitamin B, and ±0.5% lactic acid.

APPENDIX B

The Urban Scene[1]

One of the duties of an administrator of urban African affairs is to act on request as a guide to those visiting the African suburbs, officially known as "townships." Visitors range from senior politicians and leaders of industry, professional men and academicians, to housewives, holiday makers and classes of school children. Some are in search of information, many are satisfying a natural curiosity, a few are the veriest sightseers. They may come from any part of Rhodesia, Africa, or countries beyond. Often they are accompanied by their Bulawayo hosts, the great majority of whom are Europeans and, even if they live in the city, are seeing an African Township for the first time.

Their reaction is predictable; they are surprised. It is not as they expected. Tarred roads, well-built houses, neat gardens, shopping centers, clinics, cocktail bars, youth clubs, halls, playing fields, schools, libraries—the array of amenities and their quality contrast with vague notions of "shantytowns" or even with experience of great city slums. The bustle of suburbia, housewives shopping, children streaming home from school, neighbors chatting over the fence, postmen on their rounds, lack the Africanness which it was assumed would pervade the scene. It is the air of normality, the very absence of strangeness, that is strange. Often the reaction is greatest among those who have lived in Bulawayo all their lives and thought they knew what to expect. "The trouble is they never thought of Africans as *people* before," commented a City Councilor drily. With realization may come relief and the assumption that "they are all right," leading to such bizarre questions as, "Aren't they grateful for all you have done?"

Others readjust their thinking and ask questions about services and living conditions which show an awareness that they are seeing only the tip of the iceberg. Sometimes there is inclination and time to stop and talk, or call at one of the homes, where one is known and welcome. Then "Africanness" begins to emerge, but in a deeper sense. Superficial signs have always been there—English is commonly understood, but much friendly conversation is in the vernacular; babies go round on their mother's backs, prams are unusual; a loaf of bread comes home from the grocer balanced on a little girl's head; mother's sewing machine may travel in this way, too; there are children everywhere, far more than can be related by European equivalents to the number of adults seen; a goat is being led through the street, it will be slaughtered (at the abattoir, of course!) and eaten in observance of African custom; friends are sitting in the beer gar-

[1]This statement, excerpted from the report of an experienced administrator in the Housing and Amenities Department, presents one view of urban African life (35, *pp. 43–44*). Quoted with permission.

den circulating a large pot of African beer. Such indications of other ways of life intermingle with city suburban norms.

As the visitor explores, he will be confronted with a melting pot of two societies, of people who shuttle between rural subsistence and urban penury, where the illiterate mother sends her young son to read the new mathematics under a street lamp, where the emancipated and unemancipated jostle in search of a new security of house, land, or learning, of employment or of the vote. The visitor will be made aware of complexity and flux, of contrast and inconsistency, of hopes and frustrations and fears that belong to a hundred thousand differing African families in the urban areas of Rhodesia.

APPENDIX C

The Cry of My People[1]

The days of happiness have disappeared, and sorrows of all kinds have fallen on my people. They cry facing east and west, north and south. But from all sides they receive no help.

Their inheritance and their land have been taken by their rulers. Look and see how they are treated! You will feel great sorrow for them.

They are orphans in the land of their forefathers, and their fathers live like widows deprived of their property. They are not respected. The respect they had in the past has been given to the race of their rulers.

The free waters of the world that were found all over the land, and of which they drank freely in the past, are now sold to them. The firewood of the forests, which they formerly gathered without question, is now forbidden to them and presented for sale. "Oh, Wuuuuuuuuuuu!" they cry.

Their rulers chase them from every direction. They have no proper homes, for they can be moved wherever and whenever their rulers please. Their rulers have introduced separate developments which are unfairly determined. The few have more land, yet millions have only a quarter of the land compared to the few. Taxes have been levied on nearly everything that people need.

My people are now stretching their hands in all directions seeking for help, but they cannot get it. "Oh, Wuuuuuuuuuu!" they cry and cry, but still there is no help from any corner of the world.

"Our fathers made a great mistake in letting these people enter our land," they cry. The forefathers are no longer living to see how their folly has made their sons and daughters suffer. "We are always carrying the evil deeds of our rulers," they say. "The rulers, being the intruders, are ruling us with their harsh laws."

"Help, help, help!" they cry, but they do not get help. They get their food only after hard struggle, yet their rulers have plenty of food. The ruler's dogs feed on better food than my people, Because of all this they have great doubts that human beings were created equal by God. "Is God there? Is God seeing all this suffering upon us?" they ask themselves.

Their life is always in danger. Their skins have lost their sparkle because of suffering and discouragement that have no end. Both the elders

[1]This statement, written in August 1970 by a 30-year-old Ndebele, suggests something of the perceptions of one rather articulate urban African man. Like many of the personal statements quoted in previous pages, this one contains inaccuracies of fact, but the more important question is the extent to which it portrays popular opinion. As an African administrator observed, with utmost tact, the statement can at least be taken to "suggest" that not all Rhodesians are in agreement about the quality of contemporary African life. (Quoted with permission.)

and the young ones are treated the same. The young ones of their rulers receive the same respect as their fathers. They are all Bosses. The young men of my people do the hardest work and their girls are the nannies in their rulers' homes. Both the young men and young girls get only a pittance to allow them to live, yet they, the rulers rejoice and live a life of luxury.

The happy days my people enjoyed in the past have been replaced by hardships and unending sorrows. Their eyes are now blind and cannot see good or bad.

"Wuuuuuuuuuuuuuuuuuuu." They cry and cry for help which they are doubtful that they will ever receive.

—SILINDILE

BIBLIOGRAPHY

1. Ashton, H. The Basuto. London; Oxford University Press; 1952.

2. Bacon, M. K. Cross-cultural studies of drinking. Pp. 171-194. In: Bourne, P. G. and Fox, R., eds. Alcoholism; progress in research and treatment. New York; Academic; 1973.

3. Baddeley, F. African beerhalls. Thesis, University of Cape Town School of Architecture; 1966.

4. Barry, H., 3d. Sociocultural aspects of alcohol addiction. Res. Publ. Ass. nerv. ment. Dis. **46**: 455–471, 1968.

5. Beattie, J. Bunyoro, an African kingdom. New York; Holt; 1960.

6. Bothwell, T. H. Iron overload in the Bantu. In: Dreyfus, J. C. and Schapira, G., eds. Iron metabolism; an international symposium. Berlin; Springer; 1964.

7. Bothwell, T. H., Charlton, R. W. and Seftel, C. Oral iron overload. S. Afr. med. J. **32**: 892–900, 1965.

8. Bothwell, T. H. and Finch, C. A. Iron metabolism. Boston; Little, Brown; 1961.

9. Braidwood, R. J., Sauer, J. D., Helback, H., Mangelsdorf, P. C., Coon, C. S., Linton, R., Steward, J. and Oppenheim, A. L. Symposium: did man once live by beer alone? Amer. Anthrop. **55**: 515–526, 1955.

10. Brettler, P. Sorghum in the native brewing industry. Bulawayo, Rhodesia; Municipal Brewery; n.d. [Mimeographed.]

11. Bruun, K. Drinking practices and their social function. Pp. 218-228. In: Lucia, S., ed. Alcohol and civilization. New York; McGraw-Hill; 1963.

12. Buchanan, W. M. Siderosis in Rhodesian Africans. Cent. Afr. J. Med. **12**: 199–207, 1966.

13. Buchanan, W. M. Bantu siderosis, with special reference to Rhodesian Africans. (Faculty of Medicine Research Lecture Series, No. 1.) Salisbury; University College of Rhodesia; 1967.

14. Buchanan, W. M. Bantu siderosis, a review. Cent. Afr. J. Med. **15**: 105–113, 1969.

15. Bulawayo, Rhodesia. African administration in Bulawayo. Bulawayo; City Council, African Administration Department; 1957. [Mimeographed.]

16. Bulawayo, Rhodesia. Some notes on the liquor undertaking. Bulawayo; City Council, Housing and Amenities Department; [1968?] [mimeographed.]

17. Bulawayo, Rhodesia. Report by Work Study Section. (Report No. 36.) Bulawayo; African Administration Department, Liquor Selling Undertaking; 1962.

18. Bulawayo, Rhodesia. African beer sales; report to the Town Clerk. Bulawayo; 1967. [Mimeographed.]

19. Bulawayo, Rhodesia. Basic principles of African beer brewing; prepared by brewery staff (28th April). Bulawayo; 1970. [Mimeographed.]

20. Bullock, C. The Mashona and the Matabele. Cape Town; Juta; 1950.

21. BUNZEL, R. The role of Alcoholism in two Central American cultures. Psychiatry **3**: 361–387, 1940.

22. CAVAN, S. Liquor license: an ethnography of bar behavior. Chicago; Aldine; 1966.

23. CHAKAIPA, P. Dzasukwa mwana–asina–hembe. (The beer pots have been washed; the child has no clothes.) Salisbury; Longmans of Rhodesia; 1967.

24. CHILD, H. The history of the amaNdebele. Salisbury; Government of Rhodesia, Ministry of Internal Affairs; 1968.

25. CHILD, I., BACON, M. K., BARRY, H., 3d, BUCHWALD, C. and SNYDER, C. R. A cross-cultural study of drinking. Quart. J. Stud. Alc., Suppl. No. 3, 1965.

26. COLSON, E. and GLUCKMAN, M., eds. Seven tribes of British Central Africa. London; Oxford University Press; 1951.

27. DANEEL, M. L. The god of the Matopos hills; an essay on the Mwari cult in Rhodesia. Gravenhage; Mouton; 1970.

28. DEAN, G. The porphyrias: the royal purple? J. Roy. Coll. Phys., Lond. **5**: 47–61; 1970.

29. DUNN, W. L. Iron loading, fibrosis, and hepatic carcinogenesis. Arch. Path. **83**: 258–266, 1967.

30. FALLERS, L. A. Bantu bureaucracy; a study of integration and conflict in the political institutions of an East African people. Cambridge; Heffer; 1957.

31. FIELD, P. B. A new cross-cultural study of drunkenness. Pp. 48–74. In: PITTMAN, D. J. and SNYDER, C. R., eds. Society, culture and drinking patterns. New York; Wiley; 1962.

32. FLOYD, B. N. Changing patterns of African land use in Southern Rhodesia. Hum. Probl. **25**: 20–39; 1959.

33. FOSTER, G. Applied anthropology. Boston; Little, Brown; 1969.

34. GARGETT, E. S. The welfare of the urban community. In: RHODESIA. COUNCIL OF SOCIAL SERVICES. Report of the national conference. 1968. [Mimeographed.]

35. GARGETT, E. S. Welfare services in an urban African area [Bulawayo, Rhodesia]. Ph.D. dissertation, University of London; 1971.

36. GEISER, P. The contemporary Ainu, a people in search of society. Hum. Organiz. **30**: 31–38, 1971.

37. GELFAND, M. Alcoholism in contemporary African society. Cent. Afr. J. Med. **12**: 12–13, 1966.

38. GERMAN, G. A. Psychiatric syndromes. Kampala, Uganda; Makerere University Department of Psychiatry; n.d. [Mimeographed.]

39. GOULD, J. and KOLB, W. L. A dictionary of the social sciences. New York; Free Press; 1964.

40. GRAVES, T. D. Acculturation, access, and alcohol in a tri-ethnic community. Amer. Anthrop. **69**: 306–321, 1967.

41. GRAVES, T. D. The personal adjustment of Navajo Indian migrants to Denver, Colorado. Amer. Anthrop. **72**: 35–54, 1970.

42. GRAVES, T. D. Drinking and drunkenness among urban Indians. In:

Waddell, J. O. and Watson, M., eds. The American Indian in urban society. Boston; Little, Brown; 1971.

43. Gussman, B. W. African life in an urban area, a study of the African population of Bulawayo. 2 vols. Bulawayo; Federation of African Welfare Societies in Southern Rhodesia; 1952. [Mimeographed.]

44. Gutkind, P. C. W. African urbanism, mobility and the social network. In: Breese, G., ed. The city in newly developing countries; readings on urbanization. Englewood Cliffs, NJ; Prentice-Hall; 1969.

45. Gutkind, P. C. W., ed. The passing of tribal man in Africa. (International Studies in Sociological and Social Anthropology, Vol. 10.) Leiden; Brill; 1970.

46. Haberman, P. W. Denial of drinking in a household survey. Quart. J. Stud. Alc. **31**: 710–717, 1970.

47. Harwood, A. Beer-drinking and famine in a Safwa village; a case of adaptation in a time of crisis. Kampala, Uganda; Makerere Institute of Social Research; 1964. [Mimeographed.]

48. Hawthorn, H. B., Belshaw, C. S. and Jamieson, S. M. The Indians of British Columbia and alcohol. Alcsm Rev., B.C. **2** (No. 3): 10–14, 1957.

49. Heath, D. B. Peasants, revolution, and drinking; interethnic drinking patterns in two Bolivian communities. Hum. Organiz. **30**: 179–186, 1971.

50. Holleman, J. F. Some Shona tribes of Southern Rhodesia. In: Colson, E. and Gluckman, M., eds. Seven tribes of British Central Africa. London; Oxford University Press; 1951.

51. Holleman, J. F. Accommodating the spirit amongst some north-eastern Shona tribes. (Rhodes-Livingstone Papers, No. 22.) London; Oxford University Press; 1953.

52. Holleman, J. F. Chief, council, and commissioner. London; Oxford University Press; 1969.

53. Honigmann, J. J. Dynamics of drinking in an Austrian village. Ethnology **2**: 157–179, 1963.

54. Honigmann, J. J. and Honigmann, I. Drinking in an Indian–White community. Quart. J. Stud. Alc. **5**: 576–619, 1945.

55. Honigmann, J. J. and Honigmann, I. How Baffin Island Eskimo have learned to use alcohol. Social Forces **44**: 73–83, 1965.

56. Horton, D. The functions of alcohol in primitive societies; a cross-cultural study. Quart. J. Stud. Alc. **4**: 199–311, 1943.

57. Howard, J. A. Report on tour of liquor undertakings; Pretoria, the Reef and Durban; prepared by the Administrative Officer, Liquor Sales and Distribution. Bulawayo; African Housing and Amenities Department; 1970. [Mimeographed.]

58. Howard, J. A. Report on tour of liquor undertakings: Salisbury and Gwelo; prepared by the Administrative Officer, Liquor Sales and Distribution. Bulawayo; African Housing and Amenities Department; 1970. [Mimeographed.]

59. Hughes, A. J. B. Kin, caste and nation among the Rhodesian Ndebele. (Rhodes-Livingstone Papers, No. 25.) Manchester; Manchester University Press; 1956.

60. HURT, W. R. and BROWN, R. M. Social drinking patterns of the Yankton Sioux. Hum. Organiz. **24**: 222–230, 1965.

61. HUTCHINSON, B. Alcohol as a contributing factor in social disorganization. Rev. Antropol. **9**: 1–13, 1961.

62. JELLINEK, E. M. Phases of alcohol addiction. Quart. J. Stud. Alc. **13**: 673–684, 1952.

63. JELLINEK, E. M. The disease concept of alcoholism. Highland Park, NJ; Hillhouse Press; 1960.

64. JOSEPH, A. Physician and patient; some aspects of inter-personal relations between physicians and patients, with special regard to the relationship between White physicians and Indian patients. Appl. Anthrop. **1** (No. 4): 1–6, 1942.

65. KEESING, R. M. and KEESING, F. New perspectives in cultural anthropology. New York; Holt, Rinehart & Winston; 1971.

66. KIDD, D. The essential kafir. London; Adam & Charles Black; 1904.

67. KILEFF, C. Black suburbanites; adaptation to Western culture in Salisbury, Rhodesia. Ph.D. dissertation, Rice University; 1970.

68. KILEFF, C. and PENDELTON, W., eds. Urban man in southern Africa. Chattanooga; University of Tennessee; n.d. [Mimeographed.]

69. KRIGE, J. D. and KRIGE, E. J. The realm of a rain-queen. London; Oxford University Press; 1943.

70. KUPER, H. The Swazi; a South African kingdom. (Case Studies in Cultural Anthropology.) New York; Holt, Rinehart & Winston; 1963.

71. KUPER, H., HUGHES, A. J. B. and VAN VELSEN, J. The Shona and Ndebele of Southern Rhodesia. London; International African Institute; 1955.

72. LEMERT, E. M. Alcohol and the Northwest Coast Indians. Univ. Calif. Publ. Cult. Soc. **2**: 303–406, 1954.

73. LEMERT, E. M. Alcoholism; theory, problems and challenge. *III.* Alcoholism and the sociocultural situation. Quart. J. Stud. Alc. **17**: 306–317, 1956.

74. LEMERT, E. M. Alcohol, values and social control. Pp. 553–571. In: PITTMAN, D. J. and SNYDER, C. R., eds. Society, culture, and drinking patterns. New York; Wiley; 1962.

75. LEMERT, E. M. An interpretation of society's efforts to control the use of alcohol. In: Alcoholism—society's responsibility. Berkeley; California State Department of Health; 1958.

76. LEMERT, E. M. Comment on Mandelbaum's article, "Alcohol and culture." Curr. Anthrop. **6** (No. 3): 291, 1965.

77. LEVY, L. F., FORBES, J. K. and PARIRENYATWA, T. S. Epilepsy in Africans. Cent. Afr. J. Med. **10**: 241–249, 1964.

78. LINTON, R. The study of man. New York; Appleton-Century-Crofts; 1936.

79. LOMNITZ, L. Patterns of alcohol consumption among the Mapuche. Hum. Organiz. **28**: 287–296; 1969.

80. LUDWIG, A. M. On and off the wagon; reasons for drinking and abstaining by alcoholics. Quart. J. Stud. Alc. **33**: 91–96, 1972.

81. MACANDREW, C. and EDGERTON, R. B. Drunken comportment; a social explanation. Chicago; Aldine; 1969.

82. McCarthy, R. G. Drinking and intoxication; selected readings in social attitudes and controls. New Brunswick, NJ; Rutgers Center of Alcohol Studies; 1959.

83. Mandelbaum, D. G. Alcohol and culture. Curr. Anthrop. **6:** 281–293, 1965.

84. Mangin, W. Drinking among Andean Indians. Quart. J. Stud. Alc. **18:** 55–56, 1957.

85. Marconi-T., J., Varella, A., Rosenblatt, E., Solari, G., Marchesse, I., Alvarado, R. and Enriquez, W. A survey on the prevalence of alcoholism among the adult population of a suburb of Santiago. Quart. J. Stud. Alc. **16:** 438–446, 1955.

86. May J. Survey of the urban African drinking patterns in the Bulawayo municipal townships. Salisbury; University of Rhodesia, Department of Sociology, Alcoholism Research Unit; 1970. [Mimeographed.]

87. Mazobere, C. G. Doro kudya. [Beer is food.] Gwelo, Rhodesia; Mambo Press; 1964. (Printed in Shona. Also printed in an Ndebele version as: Ncube G. D. Utshwala yikudla. [1968.])

88. Mhlagazanhlansi. My friend Kumalo. Bulawayo; Rhodesia Printing & Publishing Co.; 1946.

89. Murphree, M. W. Christianity and the Shona. (Monogr. on Social Anthropology, No. 36.) London; London School of Economics; 1969.

90. Nelson, G. K., Novellie, L., Reader, D. H., Reuning, H. and Sachs, H. Psychological, nutritional, and sociological studies of Kaffir beer. Pretoria; South African Council for Scientific and Industrial Research, Johannesburg Kaffir Beer Research Project; 1964. [Mimeographed.]

91. Netting, R. M. Beer as a locus of value among the West African Kofyar. Amer. Anthrop. **66:** 375–384, 1964.

92. Nielsen, P. The Matabele at home. Bulawayo; Davis; 1913.

93. Orfei, E., Volini, F. I., Madera-Orsini, F., Minick, O. T. and Kent, G. Effect of iron loading on hepatic injury induced by ethionine, a re-evaluation. Amer. J. Path. **52:** 547–565, 1968.

94. Orley, J. H. Culture and mental illness; a study from Uganda. Nairobi; East African Publishing House; 1970.

95. Peaden, W. R. Missionary attitudes to Shona culture, 1893–1923. Salisbury; University of Rhodesia, Central African Historical Association; 1970.

96. Pittman, D. J. and Snyder, C. R., eds. Society, culture, and drinking patterns. New York; Wiley; 1962.

97. Platt, B. S. Some traditional alcoholic beverages and their importance in indigenous African communities. Proc. Nutr. Soc. **14:** 115–124, 1955.

98. Popham, R. E. and Yawney, C. D. Culture and alcohol use; a bibliography of anthropological studies. Toronto; Alcoholism and Drug Addiction Research Foundation of Ontario; 1966.

99. Rachman, I. Epilepsy in African hospital practice. Cent. Afr. J. Med. **16:** 201–204, 1970.

100. Ransford, O. N. The rulers of Rhodesia. London; Murray; 1968.

101. READER, D. H. Tribalism and detribalization in southern and central Africa. Zambezia 1: 55–75, 1970.

✓102. READER, D. H. and MAY, J. Drinking patterns in Rhodesia; Highfield African Township, Salisbury. (Department of Sociology, Occasional Paper No. 5.) Salisbury; University of Rhodesia; 1971.

103. RHODESIA. CENTRAL STATISTICAL OFFICE. Report on urban African budget survey. Salisbury; 1970.

✓104. RICHARDS, A. Land, labour and diet in Northern Rhodesia; an economic study of the Bemba tribe. London; Oxford University Press; 1939.

✓105. RICHARDS, A. Hunger and work in a savage tribe. New York; Free Press; 1948.

✓106. ROBBINS, M. C. and POLLNAC, R. B. Drinking patterns and acculturation in rural Buganda. Amer. Anthrop. **71**: 276–285, 1969.

107. ROSS, W. F. Epidemiological study of trauma in Mpilo Central Hospital, Bulawayo, Southern Rhodesia. (Proceedings of the Central African Scientific and Medical Congress, Lusaka, August 1963.) Oxford; Pergamon; 1965.

108. ROSS, W. F. and WHITE, A. Surgical services in an African community. Cent. Afr. J. Med. **10** (Suppl.): 1–21, 1964.

109. RUBINGTON, E. The chronic drunkenness offender. Ann. Amer. Acad. polit. soc. Sci. **315**: 65–72, 1958.

110. SALISBURY. LIQUOR UNDERTAKING DEPARTMENT. Rufaro, the story of African beer in Salisbury. Salisbury; 1967.

111. SALISBURY. LIQUOR UNDERTAKING DEPARTMENT. Rufaro, the story of African beer in Salisbury. Salisbury; 1969.

✓112. SANGREE, W. H. The social functions of beer drinking in Bantu Tiriki. Pp. 6–21. In: PITTMAN, D. J. and SNYDER, C. R., eds. Society, culture, and drinking patterns. New York; Wiley; 1962.

✓113. SARIOLA, S. Dryckesseder på Colombias landsbygd. (Drinking customs in rural Columbia.) Alkoholpolitik, Hels. **24**: 127–131, 1961.

114. SCHWAB, W. B. Social stratification in Gwelo. In: SOUTHALL, A., ed. Social change in modern Africa. London; Oxford University Press; 1961.

115. SCHWAB, W. B. Comparative field techniques in urban research in Africa. In: FREILICH, M., ed. Marginal natives; anthropologists at work. New York; Harper & Row; 1970.

116. SCUDDER, T. Relocation, agricultural intensification, and anthropological research. In: BROKENSHA, D. and PEARSALL, M., eds. The anthropology of development in sub-Saharan Africa. (Society for Applied Anthropology, Monogr. No. 10.) Lexington, KY; 1969.

117. SEFTEL, H. C., MALKIN, C., SCHMAMAN A., ABRAHAM, C., LYNCH, S. R., CHARLTON, R. W. and BOTHWELL, T. H. Osteoporosis, scurvy, and siderosis in Johannesburg Bantu. Brit. Med. J. **1**: 642–646, 1966.

118. SHERLOCK, S. Diseases of the liver and biliary system, 3d ed. Oxford; Bothwell; 1963.

✓119. SIMMONS, O. G. Drinking patterns and interpersonal performance in a Peruvian Mestizo community. Quart. J. Stud. Alc. **20**: 103–111, 1959.

120. SIMMONS, O. G. Ambivalence and the learning of drinking behavior in a Peruvian community. Amer. Anthrop. **62:** 1018–1027, 1960.

121. SMITH, D. K. Development and testing of a low-cost periodical for use in developing areas of the world. Ph.D. dissertation, University of Oregon; 1969.

122. SMITH, D. K. World view of the Matabele. Bulawayo; Daystar; n.d. [Mimeographed.]

123. SNYDER, C. R. Alcohol and the Jews; a cultural study of drinking and sobriety. (Rutgers Center of Alcohol Studies, Monogr. No. 1.) New Brunswick, NJ; 1958.

124. SOMMER, R. Designed for drinking. In: Personal space; the behavioral basis of design. Englewood Cliffs, NJ; Prentice-Hall; 1969.

125. SOUTHALL, A. W. and GUTKIND, P. C. W. Townsmen in the making—Kampala and its suburbs. Kampala, Uganda; East African Institute of Social Research; 1957.

126. SPRADLEY, J. P. You owe yourself a drunk; an ethnography of urban nomads. Boston; Little, Brown; 1970.

127. STOPFORTH, P. Survey of Highfield African township. (Department of Sociology, Occasional Paper No. 6.) Salisbury; University of Rhodesia; 1971.

128. ULLMAN, A. D. First drinking experience as related to age and sex. Pp. 259–266. In: PITTMAN, D. J. and SNYDER, C. R., eds. Society, culture, and drinking patterns. New York; Wiley; 1962.

129. UMUNNA, I. The drinking culture of a Nigerian community: Onitsha. Quart. J. Stud. Alc. **28:** 529–537, 1967.

130. UNIVERSITY COLLEGE OF RHODESIA, INSTITUTE FOR SOCIAL RESEARCH, ALCOHOLISM RESEARCH UNIT. Accounts from depth interviews. [Unpublished field notes, 1969.]

131. UNIVERSITY COLLEGE OF RHODESIA, INSTITUTE FOR SOCIAL RESEARCH, ALCOHOLISM RESEARCH UNIT. Survey of urban African drinking patterns, Bulawayo. [Interview schedule and unpublished field data, 1970.]

132. WALLACE, A. F. C. Administrative forms of social organization. Module, Reading, MA **9:** 1–12, 1971.

133. WALLACE, S. E. Skid Row as a way of life. Totowa, NJ; Bedminster; 1965.

134. WASHBURNE, C. Primitive drinking; a study of the uses and functions of alcohol in preliterate societies. New York; College & University Press; 1961.

135. WEINRICH, A. K. H. (Sister Mary Aquina, O.P.) Chiefs and councils in Rhodesia. Columbia; University of South Carolina Press; 1971.

136. WHITE, A. Penetrating wounds of the chest. E. Afr. Med. J. **40:** 522–533, 1963.

137. WITZLEBEN, C. L. and CHAFFEY, N. J. Response of the iron-loaded liver to prolonged ethionine feeding. Arch. Path. **80:** 447–452, 1965.

138. WOLCOTT, H. F. A Kwakiutl village and school. New York; Holt, Rinehart & Winston; 1967.

139. WOLCOTT, H. F. An ethnographic approach to the study of school administrators. Hum. Organiz. **29**: 115–122, 1970.
140. WOLCOTT, H. F. Too true to be good; observations on the subculture of American missionaries in urban Africa. Pract. Anthrop. **19**: 241-258, 1972.
141. WOLCOTT, H. F. The man in the principal's office: an ethnography. New York; Holt, Rinehart & Winston; 1973.
142. WOLCOTT, H. F. Feedback influences on fieldwork, or, a funny thing happened on the way to the beer garden. In: KILEFF, C. and PENDELTON, W., eds. Urban man in southern Africa. [Submitted for publication.]

INDEX OF NAMES

INDEX OF SUBJECTS